AA003302

Transport and Energy Processes 2014

Core Programming Area at the 2014 AIChE Annual Meeting

Atlanta, Georgia, USA
16-21 November 2014

ISBN: 978-1-5108-1275-8

Printed from e-media with permission by:

Curran Associates, Inc.
57 Morehouse Lane
Red Hook, NY 12571

proceedings.com

Some format issues inherent in the e-media version may also appear in this print version.

Copyright© (2014) by AIChE
All rights reserved.

Printed by Curran Associates, Inc. (2015)

For permission requests, please contact AIChE
at the address below.

AIChE
120 Wall Street, FL 23
New York, NY 10005-4020

Phone: (800) 242-4363
Fax: (203) 775-5177

www.aiche.org

Additional copies of this publication are available from:

Curran Associates, Inc.
57 Morehouse Lane
Red Hook, NY 12571 USA
Phone: 845-758-0400
Fax: 845-758-2634
Email: curran@proceedings.com
Web: www.proceedings.com

TABLE OF CONTENTS

389140 Multi–Scale Model for the Rational Design of Fischer–Tropsch Catalysts

Abstract:

For the conversion of syngas to hydrocarbons, Fischer and Tropsch proposed a surface carbide mechanism, which involves direct CO dissociation and subsequent hydrogenation of chemisorbed carbon (C) to form CH_x monomers that initiate and grow hydrocarbon chains [1-4]. Meanwhile, Storch and coworkers suggested that condensation of oxygen–containing intermediates (HCOH, hydroxycarbene) was responsible for C–C bond formation [5, 6]. Although most of these reaction schemes have been studied over the last decades, there are conflicting views concerning the actual reaction intermediates as well as the correct sequence of intermediate and termination steps. This situation reflects the experimental difficulties of investigating the reactions for the Fischer–Tropsch (FT) mechanism. Molecular modeling has been proven to be a helpful tool to complement and guide experimental studies. In order to anticipate the initiation and termination steps associated with the conversion of syngas ($CO+H_2$) to liquid hydrocarbons through FT synthesis, simulations were carried out employing theoretical methods such as Density Functional Theory (DFT) and Kinetic Monte Carlo (KMC); using Co and Co–based structures as the prototype catalysts. Consequently, using molecular simulations in combination with experiments lowers the costs related to design as the simulations can be used to screen only the useful catalysts that can be subjected to experiment. Moreover, the elucidation of this process provides an improved understanding of the molecular mechanisms of hydrocarbon formation from syngas.

[1] Van Santen RA, van Leeuwen PWNM, Moulijn JA. Catalysis: An Integrated Approach: Elsevier; 2000.

[2] Aho A, Antonietti M, Arndt S, Behrens M, Bill E, Brandner A, et al. Chemical Energy Storage: De Gruyter; 2012.

[3] Steynberg A, Dry M. Fischer-Tropsch Technology: Elsevier Science; 2004.

[4] Gates BC, Knoezinger H, Jentoft FC. Advances in Catalysis: Elsevier Science; 2011.

[5] Anderson RB, Kölbel H, Rálek M. The Fischer-Tropsch synthesis: Academic Press; 1984.

[6] Ojeda M, Nabar R, Nilekar AU, Ishikawa A, Mavrikakis M, Iglesia E. CO activation pathways and the mechanism of Fischer–Tropsch synthesis. Journal of Catalysis. 2010;272:287-97.

Copyright © American Institute of Chemical Engineers. All rights reserved.

370610 Characterization and Thermal Stability of Oxidized Sn/Pt(100) Alloy Surfaces

Abstract:

Pt-Sn surface alloys formed on a Pt(100) substrate are remarkably stable to oxidation - no thermal dissociative adsorption of O_2 occurs under UHV conditions. Nonetheless oxidation of these alloys can be carried out clearly and controllably in UHV by using O_3 (ozone). Ozone is a reactive oxidant, creating high "effective" oxygen pressure in UHV such that large concentrations (up to Q_O = 2.2 and 2.1 ML) of chemisorbed and "oxidic" oxygen were produced on the two ordered c(2x2) and (3(rt2xrt2))R45 degrees Sn/Pt(100) surface alloys, respectively. AES, LEED, HR-XPS, and TPD were used to characterize the oxidized alloy surface, measure the oxygen uptake kinetics, and probe the thermal stability of the oxidized surface. Ozone exposed on the alloy surfaces at 300 K oxidized Sn. Annealing the alloy surfaces during TPD revealed two types of O_2 desorption peaks on each alloy surface: peaks observed at 910 and 1100 K on the c(2x2) Sn/Pt(100) surface alloy are attributed to Pt-O-Sn and Sn-O interactions respectively, and peaks at 750 and 1100 K observed on the (3(rt2xrt2))R45degrees Sn/Pt(100) surface alloy are attributed to Pt-O and Sn-O interactions. At both surfaces Sn-O interactions lead to the formation of SnO_x at high temperature (> 800 K), which decomposed around 1100 K. SnO_x formed on these surfaces is destabilized by at least 100 K due to the presence of Pt as compared to the oxide formed on a thick layer of Sn on the Pt(100) surface. In contrast, PtO_x on the (3(rt2xrt2))R45degrees surface is 90 K more stable in the presence of SnO_x on the surface as compared to the oxide formed on the clean Pt(100) surface. Both Sn/Pt(100) surface alloys oxidized more readily than Pt(100), and a c(2x2) Sn overlayer was even more reactive for dissociating ozone.

Copyright © American Institute of Chemical Engineers. All rights reserved.

392403 New Mesoporous ZSM5 Synthesis and Its Use in Hydrogenation of Naphthalene in the Absence and Presence of Sulfur

Abstract:

Deep hydrogenation of aromatics is an important reaction in modern oil refinery. The key of an aromatic hydrogenation unit is high active, sulfur resistant hydrogenation catalyst. In this paper, the catalytic performance of a newly developed hydrogenation catalyst, platinum supported on mesoporous ZSM-5 (Pt/MZ-5), was tested in naphthalene (model compound of light cycle oil) hydrogenation. Pt/ZSM-5 and Pt/Alumina were used as reference. In some experiments, dibenzothiophene (DBT) was added to test the sulfur tolerance of catalysts. In the absence of DBT, Pt/MZ-5 and Pt/Alumina showed much higher hydrogenation activities and decalin selectivity than Pt/ZSM-5. The selectivity to decalin exceeds 97% over Pt/MZ-5 at temperature range between 523 and 573 K. Furthermore, Pt/MZ-5 shows better sulfur resistance than that of Pt/Alumina. In the presence of 3000 ppm dibenzothiophene, the selectivity of decalin over Pt/MZ-5 was only slightly dropped to 89.4% at 573 K, which is much higher than that of $Pt/Al2O3$ (51.4%). The excellent catalytic performance and sulfur tolerance of Pt/MZ-5 were attributed to the combination of high acidity and mesoporous structure in MZ-5.

Copyright © American Institute of Chemical Engineers. All rights reserved.

362179 Low Temperature Catalytic Hydrolysis of Carbonyl Sulfide (COS) during Hydrogen Production for Fuel Cell Applications

Abstract:

After hydrocarbon fuel is reformed (NG, LPG) or gasified (Liquid fuels), sulfur appears primarily as H_2S (10 ppmv for regulated diesel or gas – 10,000 ppmv from an oxygen blown coal gasifier) while presence of Carbonyl sulfide (COS) depends on operating conditions of the desulfurizer & an equilibrium between H_2S and CO/CO_2. Post reformer desulfurization is an important unit operation in fuel processing of these hydrocarbons during H_2 production for PEMFCs operated at low temperature (<100°C). Efficient H_2S removal can be achieved via highly dispersed ZnO supported on porous silica. COS is a relatively neutral molecule compared to highly acidic H_2S and ZnO based adsorbents suffer from almost no capacity for COS at ambient conditions. Another concern is that presence of COS is known to cause compliance-testing failures while using LPG as a raw material for H_2 production. Hydrolysis of COS to H_2S followed by H_2S adsorption is an attractive means to control the sulfur content of reformed/gasified fuel. Most studies are focused on hydrolysis at temperatures in excess of 200°C and COS concentration in excess of 5000 ppmv. The current work is, however, focused on COS removal via catalytic hydrolysis with 200 ppmv COS fed to the reactor operating at ambient (20°C) to mild temperatures (150 °C). These conditions were found to be close to industrially relevant conditions (~150 ppm & 30-60°C). As a result, there is need to develop novel materials to overcome the issue of initial catalyst activation and dissociative chemisorption responsible for active site blockage at ambient conditions. A series of catalysts consisting of Al_2O_3, TiO_2 and SiO_2 impregnated with promoter ions such as transition metals (Cu, Mn, Zn, Fe), alkaline metals (Na, K, Li), alkali earth (Mg, Ca, Ba), rare earth metals (Y, Gd, Nd, La) and Layered Double Hydroxides were tested in terms of their specific activity for COS hydrolysis at ambient conditions. GC-PFPD was used to estimate the H_2S and COS concentration simultaneously at the reactor outlet. Effect of COS concentration (200-1000 ppmv) and H_2O concentration in feed on hydrolysis rate are also discussed. Finally, in order to understand the deactivation mechanism of these catalysts the performance is monitored in presence of 10 ppmv H_2S in feed stream.

Copyright © American Institute of Chemical Engineers. All rights reserved.

374011 Conversion of Non-Transportable Brown Coals to Hydrogen By Steam Gasification Process

Abstract:

Australian brown coals are non-transportable lignite because of their high content of water. On drying, brown coals show self-ignition characteristics making them hazardous substances to transport. Their use is only limited to local thermal power generation plants. Recently, there have been advances to develop technologies to transport these coals to the point of use such as converting to char at place of production. However, in addition to environmental reasons, from economy development point of view of brown coal producing country and the energy requirement point of view of energy resources deficient country, converting brown coal to hydrogen/liquid fuels is the best suited approach. Efficiently converting brown coal to hydrogen/synthesis gas will be the main driver for this business model to be economically successful. Therefore, development of efficient conversion technologies is necessary. Currently, conventional oxygen blown gasification process is a primary commercial process to gasify coals to synthesis gas or hydrogen. However, low temperature catalytic coal gasification is more efficient and economical way to convert coal into hydrogen/synthesis gas. AIST is developing a catalytic gasification technology to efficiently convert non-transportable fossil fuels like brown coals, biomass and bio-process wastes to hydrogen/liquid fuels via synthesis gas-FT combined process. Gas composition of synthesis gas is primarily determined by temperature (thermodynamic equilibrium) and gasifying agent composition. Gasification in steam at low temperature leads to hydrogen rich gas production but chemical kinetics becomes the limiting factor for process development. This disadvantage can be overcome by using alkali catalyst. Use of catalyst for coal gasification has not been preferred because of loss of catalyst by irreversible reaction with mineral matters present in coals which means large amount of make-up catalyst will be needed. However, brown coals have less than 1 wt% of mineral matter making the make-up catalyst almost equal to operating loss of catalyst. Therefore, conversion of brown coal by catalytic steam gasification is most economical and efficient way. This study presents a brown coal catalytic gasification process to produce hydrogen rich gas or synthesis gas by simply adjusting the gasifying agent composition. The highlight of the process is the production of the hydrogen/synthesis gas at 650~700°C using steam/CO_2 as gasifying agent. Synthesis gas suitable for FT-synthesis process was produced by changing the steam/CO_2 ratio of the gasifying agent. When gasifying agent is pure steam, hydrogen and CO_2 are the main component of the gas from which hydrogen can be separated. Hydrogen can be transported by hydrogen-carrier process which involves hydrogenating the hydrogen carrier at the place of production, transporting it to consumption site and then dehydrogenating to recover the hydrogen or as liquefied H_2. FT-process suitable synthesis gas can be produced by using steam/CO_2 mixture as gasifying agent. The process does not use oxygen, thus saving significant energy needed to run ASU, and leading to higher thermal conversion efficiencies than the conventional oxygen blown gasification process.

Copyright © American Institute of Chemical Engineers. All rights reserved.

Effects of Biodiesel Contamination on Oxidative and Storage Stability of Neat and Blended Hydroprocessed Renewable Diesel

The US military is the largest consumers of middle distillate fuels in America. Many schemes have been proposed to decrease the military reliance on fossil resources by utilization of renewable sources. Second generation renewable fuels synthesized through hydroprocessing renewable feedstocks has many advantages over first generation renewable fuels, such as bioethanol and biodiesel. Hydroprocessed renewable diesels (HRD) are currently in production for large-scale testing by the U.S. Navy, and has been demonstrated as a compatible replacement or blending stock with marine diesel. The usage of biodiesel has been prohibited by the U.S. Navy in all ships and tactical platforms. Biodiesel, however, is still used as blending stock for Navy ground transportation. Concern has been raised about the potential biodiesel contamination of marine diesel, HRD, and marine/HRD blends from the use of common fuel handling systems. The present study investigates the influence of biodiesel contamination on conventional and renewable marine diesels, i.e. NATO F-76 and HRD-76. The physical properties and chemical composition of HRD-76 and biodiesel were measured. Long-term (ASTM D4625) and accelerated (ASTM D5304) test methods were used to investigate the influence of biodiesel on storage stability of HRD-76, F-76 and their blends and on fuel physical properties after long-term storage. In addition, ASTM D2274 test was conducted to study the effect of biodiesel contamination on oxidation stability of neat and blended fuels. Extended ASTM D2274 tests were also conducted to investigate the influence of long-term oxidation on fuel physicochemical properties. The oxidation precursors of fuel samples were identified using GC/MS.

381011 Sludge Treatment System Combined with Composting and Carbonization

Abstract:

SLUDGE TREATMENT SYSTEM COMBINED

WITH C AND CAEBONIZATION

N. Kobayashi[1]*, Y. Itaya[1], H. Hamabe Hisashi[2]

[1]*Department of Mechanical Engineering, Gifu University

Yanagido 1-1, Gifu-shi, Gifu, 501-1193, Japan

[2]Obu Unity Co., Ltd.

2 Kitasaki-cho Komaba 88 Obu-shi, Aich, 474-0001, Japan

*Corresponding author: Tel. +52-58-293-2621, E-mail:nsuke@gifu-u.ac.jp

Abstract: In order to reduce energy consumption during the sludge treatment process, sludge treatment combined with composting (bio-drying) and carbonization was carried out (See Figure 1). Moisture content of the sludge reduced less than 30% thorough the composting process without adding any fuel, and the addition of sludge char to compost increased the bio-drying rate significantly. A permitted range of initial sludge moisture content was 78% - 81% at the carbonization temperature of 673K - 1173K to sustain self-energy sludge treatment system.

Keywords: Sludge, Bio-drying, Sludge Composting, Carbonization

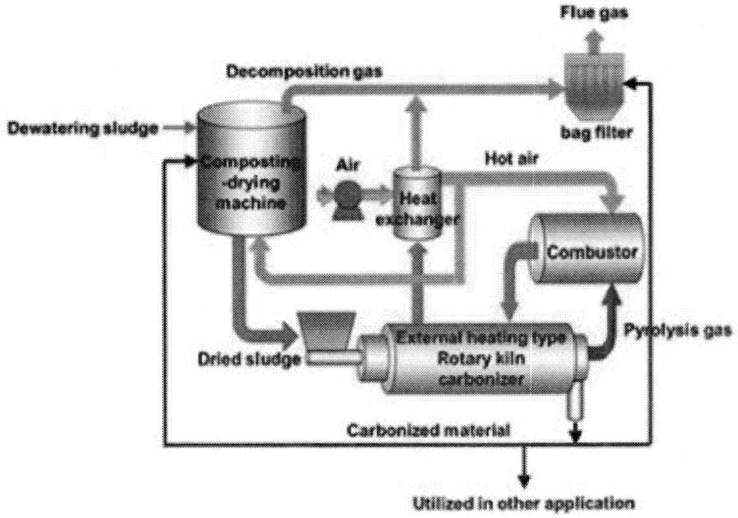

Fig. 1 New Sludge treatment system combined composting and carbonization

Copyright © American Institute of Chemical Engineers. All rights reserved.

University of South Alabama

Modeling and Analysis of Ethanol Reformer for Fuel Cell Applications

Ghanem Laith Sabeeh

Department of Chemical & Biomolecular Engineering

Overview

Department of Chemical & Biomolecular Engineering

- Alternative Fuel Source
- Steam Reforming of Ethanol
- Objective
- Modeling
- Results
- Significance of the Work
- Conclusions

Statistics about USA

Department of Chemical & Biomolecular Engineering

- 2nd largest energy consumer in the world (2010).
- 7th largest in energy consumption per capita (2010).
- 82.1% of our energy consumption comes from fossil fuels (2012).
- 1st largest in the total cumulative CO_2 emissions into the air, which is 26.8% of the total (2012).

Source: EIA (U.S. Energy Information Administration) reviewed April 12, 2014; EIA, Monthly Energy Reviewed, April 2013; ABC News

Why Alternative Fuel?

Department of Chemical & Biomolecular Engineering

Increase the efficiency of energy sources.

Decease our energy consumption.

Decrease our dependence on fossil fuels.

Why Alternative Fuel?

Department of Chemical & Biomolecular Engineering

Reduce our impact on the environments.

Reduce our reliance on foreign oil.

Hydrogen Fuel Cell

Department of Chemical & Biomolecular Engineering

- Hydrogen
  - Readily available
  - Clean and green
  - No-toxic and pollution free
  - Energy security
  - Economic growth
- Fuel Cell
  - Low emission
  - Quiet operation
  - Onsite energy generation
  - No waste

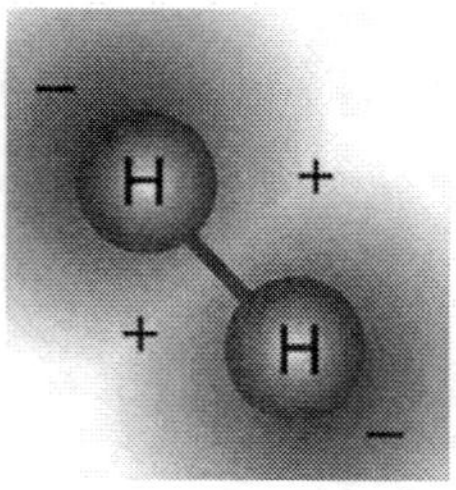

How Does Fuel Cell Works?

Department of Chemical & Biomolecular Engineering

- Fuel cell is a device that works on the principle of utilizing chemical energy from hydrogen and oxygen and converting it to electrical energy.

- Every fuel cell consists of an anode, an electrolyte and a cathode.
- Anode is in contact with hydrogen supply and cathode is in contact with oxygen supply.
- Electrolytes differ depending on the type of fuel cell.

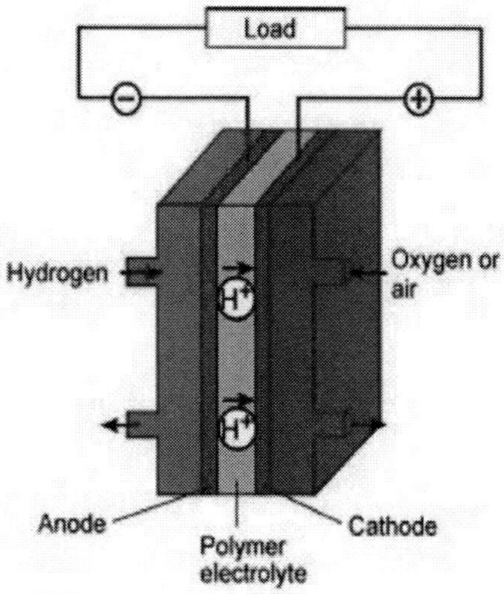

How to Provide Hydrogen?

Department of Chemical & Biomolecular Engineering

- Steam Reforming
  - Reaction of hydrocarbon with steam in presence of catalyst.
  - Endothermic reaction.
  - The burning value of the fuel is increased.
  - more efficient process which can reach 100% conversion of hydrocarbon.

Hydrocarbon Candidate

Department of Chemical & Biomolecular Engineering

- Ethanol
  - Sulfur-free compound.
  - No need for pre-reforming.
  - High hydrogen-to-carbon ratio yielding.
  - Commercially available.
  - High energy density.
  - Easy to store.
  - Safe to handle and transport.
  - Low toxicity and volatility.

Ethanol Sources

Department of Chemical & Biomolecular Engineering

- Non-renewable sources
  - Petroleum.
  - Natural gas.
  - Coal.
- Renewable sources
  - Biomass.

Objective

Department of Chemical & Biomolecular Engineering

- Conduct an unsteady state & non-isothermal steam reformer model
- Couple the reformer model with a fuel cells stack model.
- Design a reformer that supply with the collaboration of the fuel cells stack a sufficient power to a 3 bedrooms house.
- Size a back-up battery to supply the power shortage or to store excess power.

System Description

Department of Chemical & Biomolecular Engineering

- System flow sheet:

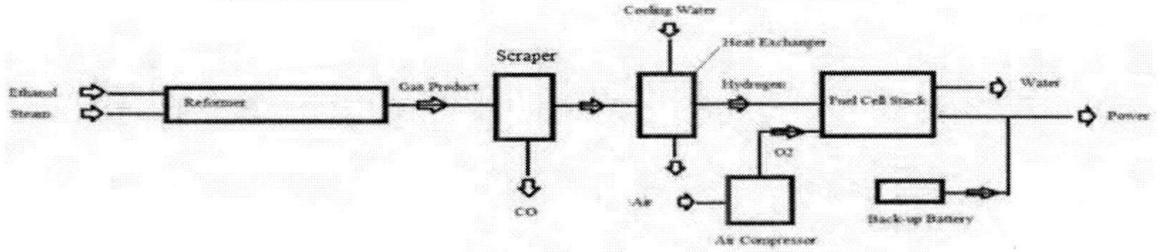

  - Carbon dioxide and methane are assumed to be removed before entering the fuel cell stack.
  - Air compressor, scrapper, and heat exchanger will not be modeled.

Kinetic Model & Parameters

- Reaction:

$$C_2H_5OH + 3H_2O \rightarrow 2CO_2 + 6H_2$$

- Reaction in the Jacket:

$$C_2H_5OH + 3O_2 + 14.286N_2 \rightarrow 2CO_2 + 3H_2O + 14.286N_2$$

- Reaction Rate (*Soyal-Baltacıoğlu, et al. 2008*):

$$r = \left(k_0 * e^{\frac{-E_A}{RT}}\right) P_e{}^{\alpha} P_w{}^{\beta}$$

  - $k_0 = 9.23 \frac{mol}{g\ cat * hr * atm^{0.92}}$
  - $E_A = 59.3 \frac{kJ}{mol}$
  - $\alpha = 1.01$
  - $\beta = -0.09$

Assumptions

- The reformer is a packed bed reactor in a tubular geometry with constant cross sectional area.
- The temperature of the wall is assumed to be the adiabatic flame temperature of burning ethanol.
- The reformer initially has no reactants.
- The reformer is initially at room temperature.
- The catalyst is spherical in shape.

Mass Balance

Department of Chemical & Biomolecular Engineering

$$\frac{\partial C_i}{\partial t} = D_{ms}\left(\frac{\partial^2 C_i}{\partial r^2} + \frac{1}{r}\frac{\partial C_i}{\partial r} + \frac{\partial^2 C_i}{\partial z^2}\right) - v_z \frac{\partial C_i}{\partial z} + \nu_i\, r$$

$C_i(z,0) = 0$ $\quad C_r(z=0,t) = C_{rin} \quad C_p(z=0,t) = 0$

$\frac{dC_i}{dz}(z=L,t) = 0 \quad \frac{dC_i}{dr}(r=0,t) = 0$

Valid for $\frac{L}{D} < 10$ and $\frac{L}{D_P} < 200$

(Hoang & Chang, 2004)

Mass Balance

Department of Chemical & Biomolecular Engineering

$$\frac{\partial C_i}{\partial t} = D_{ms}\left(\cancel{\frac{\partial^2 C_i}{\partial r^2}} + \cancel{\frac{1}{r}\frac{\partial C_i}{\partial r}} + \frac{\partial^2 C_i}{\partial z^2}\right) - v_z \frac{\partial C_i}{\partial z} + \nu_i\, r$$

$C_i(z,0) = 0$ $\quad C_r(z=0,t) = C_{rin} \quad C_p(z=0,t) = 0$

$\frac{dC_i}{dz}(z=L,t) = 0$

Valid for $\frac{L}{D} > 10$ and $\frac{L}{D_P} < 200$

(Zahedi nezhad, et al. 2009)

Mass Balance

Department of Chemical & Biomolecular Engineering

$$\frac{\partial C_i}{\partial t} = D_{ms}\left(\frac{\partial^2 C_i}{\partial r^2} + \frac{1}{r}\frac{\partial C_i}{\partial r} + \frac{\partial^2 C_i}{\partial z^2}\right) - v_z \frac{\partial C_i}{\partial z} + v_i\, r$$

$C_i(z, 0) = 0$ $\quad C_r(z = 0, t) = C_{rin}$ $\quad C_p(z = 0, t) = 0$

Valid for $\frac{L}{D} > 10$ and $\frac{L}{D_P} > 200$

(Chang & Wang, 2004)

Energy Balance

Department of Chemical & Biomolecular Engineering

$$U\,a\,(T_{Jacket} - T) - v_z(T - T_{ref})\sum_{i=0}^{n}\left(Cp_i \frac{\partial C_i}{\partial z}\right) - (T - T_{ref})\sum_{i=0}^{n}\left(Cp_i \frac{\partial C_i}{\partial t}\right) - \sum_{i=0}^{n}(C_i\, Cp_i) v_z \frac{\partial T}{\partial z}$$
$$- \Delta H_{rxn}\, r + \sum_{i=0}^{n}(C_i\, Cp_i) D_{hs}\left(\frac{\partial^2 T}{\partial r^2} + \frac{1}{r}\frac{\partial T}{\partial r} + \frac{\partial^2 T}{\partial z^2}\right) = \sum_{i=0}^{n}(C_i\, Cp_i)\frac{\partial T}{\partial t}$$

$T(z, 0) = 298.15\,K$ $\quad \mathrm{T}(z = 0, t) = T_{in}$

$\frac{dT}{dz}(z = L, t) = 0$ $\quad \frac{dT}{dr}(r = 0, t) = 0$

Valid for $\frac{L}{D} < 10$ and $\frac{L}{D_P} < 200$

(Hoang & Chang, 2004)

Energy Balance

Department of Chemical & Biomolecular Engineering

$$U\,a\,(T_{Jacket}-T) - v_z(T-T_{ref})\sum_{i=0}^{n}\left(Cp_i\frac{\partial C_i}{\partial z}\right) - (T-T_{ref})\sum_{i=0}^{n}\left(Cp_i\frac{\partial C_i}{\partial t}\right) - \sum_{i=0}^{n}(C_i\,Cp_i)v_z\frac{\partial T}{\partial z} - \Delta H_{rxn}\,r + \sum_{i=0}^{n}(C_i\,Cp_i)D_{hs}\left(\frac{\partial^2 T}{\partial r^2} + \frac{1}{r}\frac{\partial T}{\partial r} + \frac{\partial^2 T}{\partial z^2}\right) = \sum_{i=0}^{n}(C_i\,Cp_i)\frac{\partial T}{\partial t}$$

$T(z,0) = 0$ $\quad$ $T(z=0,t) = C_{rin}$

$$\frac{dT}{dz}(z=L,t) = 0$$

Valid for $\frac{L}{D} > 10$ and $\frac{L}{D_P} < 200$

(Zahedi nezhad, et al. 2009)

Energy Balance

Department of Chemical & Biomolecular Engineering

$$U\,a\,(T_{Jacket}-T) - v_z(T-T_{ref})\sum_{i=0}^{n}\left(Cp_i\frac{\partial C_i}{\partial z}\right) - (T-T_{ref})\sum_{i=0}^{n}\left(Cp_i\frac{\partial C_i}{\partial t}\right) - \sum_{i=0}^{n}(C_i\,Cp_i)v_z\frac{\partial T}{\partial z} - \Delta H_{rxn}\,r + \sum_{i=0}^{n}(C_i\,Cp_i)D_{hs}\left(\frac{\partial^2 T}{\partial r^2} + \frac{1}{r}\frac{\partial T}{\partial r} + \frac{\partial^2 T}{\partial z^2}\right) = \sum_{i=0}^{n}(C_i\,Cp_i)\frac{\partial T}{\partial t}$$

$T(z,0) = 0$ $\quad$ $T(z=0,t) = C_{rin}$

Valid for $\frac{L}{D} > 10$ and $\frac{L}{D_P} > 200$

(Chang & Wang, 2004)

Properties Estimation

Department of Chemical & Biomolecular Engineering

- Density of Mixture (kg/m^3):

$$\rho_m = 1.107 - 1.779\ 10^{-3}\ T + 9.12\ 10^{-7}\ T^2$$

- Viscosity of Mixture (Pa s):

$$\mu_m = 2.321\ 10^{-7} + 3.273\ 10^{-8}\ T - 1.964\ 10^{-12}\ T^2$$

- Thermal Conductivity of Mixture (W/m/K):

$$k_m = 0.022 + 1.401\ 10^{-4}\ T + 2.357\ 10^{-8}\ T^2$$

Diffusion & Dispersion Coefficient Estimation

Department of Chemical & Biomolecular Engineering

- Chapman–Enskog theory:

$$D_{ew} = 5.9543\ 10^{-4}\ \frac{\sqrt{T^3\left(\frac{1}{M_e}+\frac{1}{M_w}\right)}}{P\ \sigma_{ew}{}^2 \Omega_{ew}}$$

- Effective Diffusion Coefficient:

$$D_e = \frac{\phi}{\tau} D_{ew}$$

- Edwards and Richardson (1968) - The Axial Mass Dispersion Coefficient:

$$D_{ms} = \frac{v_z\ D_p}{\frac{0.73\ \phi}{Re\ Sc} + \frac{0.5}{1 + \frac{9.7\ \phi}{Re\ Sc}}}$$

- Wen and Fan (1975) - The Axial Heat Dispersion Coefficient:

$$D_{hs} = \frac{v_z\ D_p}{\frac{0.3}{Re\ Sc} + \frac{0.5}{1 + \frac{3.8}{Re\ Sc}}}$$

Overall Heat Transfer Coefficient Estimation

Department of Chemical & Biomolecular Engineering

- Sieder-Tate Equation:

$$h_o = \frac{k_t}{D_o}(1.86\, Re^{\frac{1}{3}}\, Pr^{\frac{1}{3}} \left(\frac{D_o}{L}\right)^{\frac{1}{3}} \left(\frac{\mu_m}{\mu_w}\right)^{\frac{1}{7}})$$

- Grace (1953) - The Bed Resistor:

$$h_{bed} = \frac{8\, k_e}{D_i} \qquad k_e = k_m(5 + 0.1\, Re\, Pr)$$

- Murkes & Carlsson (1988) - The Wall Resistor:

$$h_{wall} = 1.94\, Re^{0.5}\, Pr^{0.33}$$

- Overall heat transfer coefficient:

$$U = \frac{1}{\frac{1}{h_i}\frac{D_o}{D_i} + \frac{x_w}{k_t}\frac{D_o}{\left(\frac{D_o - D_i}{\ln(\frac{D_o}{D_i})}\right)} + \frac{1}{h_o}} \qquad \frac{1}{h_i} = \frac{1}{h_{wall}} + \frac{1}{h_{bed}}$$

Fuel Cells Stack Variables

Department of Chemical & Biomolecular Engineering

- The Current:

$$I = \frac{2\, F\, \varepsilon\, F_{H2}}{n_{cell}}$$

- The Voltage of The Stack:

$$V_{stack} = 6.963\, 10^{-8}\, I^4 - 2.689\, 10^{-5}\, I^3 + 3.611\, 10^{-3}\, I^2 - 0.245\, I + 35.217$$

- The Power Output:

$$P_{stack} = I\, V_{stack}(I)$$

Back-up Battery Sizing

Department of Chemical & Biomolecular Engineering

- Lead Acid rechargeable battery will be used with a voltage of 12 V.
- Convert the power lag profile into amps-hr profile.
- Chose a maximum point from the power lag profile with the duration that it takes.
- Account for the aging factor and Peukart effect.

$$C_{battery} = C_{max}\, f_a\, f_P$$

Effectiveness Factor Estimation

Department of Chemical & Biomolecular Engineering

- Weisz and Hicks (1962) - The Non-isothermal Effect of The Particle on The Effectiveness Factor:

$$k = \rho_{cat}\, k_0\, (R\,T)^{0.92} e^{\frac{-E_A}{R\,T}}$$

$$\Theta = \frac{V_p}{A_p}\sqrt{\frac{k}{D_e}}$$

$$\gamma = \frac{E_A}{R\,T_s}$$

$$\beta = \frac{-\Delta H_{rxn} D_{ew}\, C_{As}}{k\,T_s}$$

$$I\,(\beta,\gamma) = \left(2\int_0^1 c\, e^{\frac{\gamma\,\beta\,(1-c)}{1+\beta(1-c)}} dc\right)^{0.5}$$

$$\Theta_e = \frac{\Theta}{I\,(\beta,\gamma)}$$

$$\eta = \frac{1}{\Theta_e}\left(\frac{1}{\tanh(3\,\Theta_e)} - \frac{1}{3\,\Theta_e}\right)$$

Effect of Effectiveness Factor

Department of Chemical & Biomolecular Engineering

Table 1: Values of effectiveness factors for different particle diameters where the non-isothermal particle effect is considered.

Particle Diameter (mm)	Effectiveness Factor (%)
0.1	100
0.5	100
0.8	100
1	100
5	99.5
8	98.8
10	98.1
100	45.1

Effect of Pressure Drop

Department of Chemical & Biomolecular Engineering

Table 2: Parameter for the pressure drop simulation.

Name	Value	Unit
Inlet Ethanol Flow	0.0001	mol/s
Steam to Ethanol Ratio	3	
Inlet temperature	1000	K
Inlet Pressure	3	atm
Heat Supplied by the Jacket	50,000	W/m^3
Void Fraction	0.3	
Length of the Reformer	2	m
Inner Diameter of the Reformer	0.2	m

Effect of Pressure Drop

Department of Chemical & Biomolecular Engineering

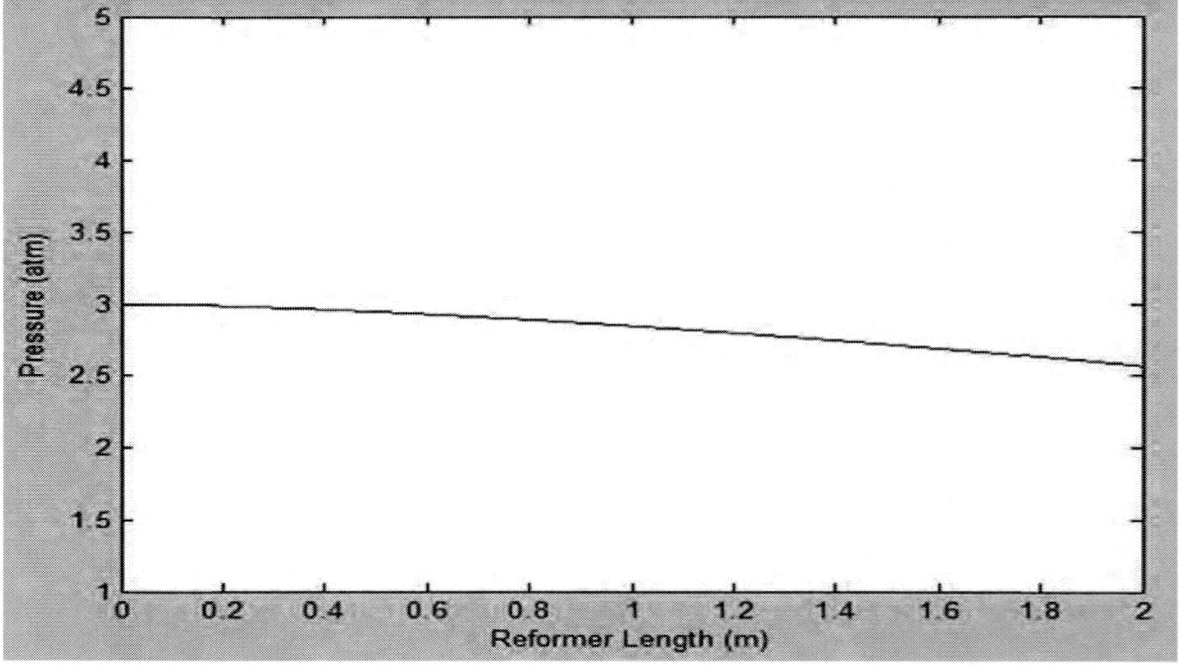

Model Validation

Department of Chemical & Biomolecular Engineering

Table 3: Values of parameters for model validation calculations.

Name	Symbol	Value	Unit
Length of Reformer	L	50	cm
Inner Diameter of Reformer	D_i	4	mm
Void Fraction	ϕ	0.3	
Steam to Ethanol Ratio	r_s	6	
Diameter of Particle	D_p	0.3	mm

Soyal-Baltacıoğlu, *et al., 2008*

Model Validation

Department of Chemical & Biomolecular Engineering

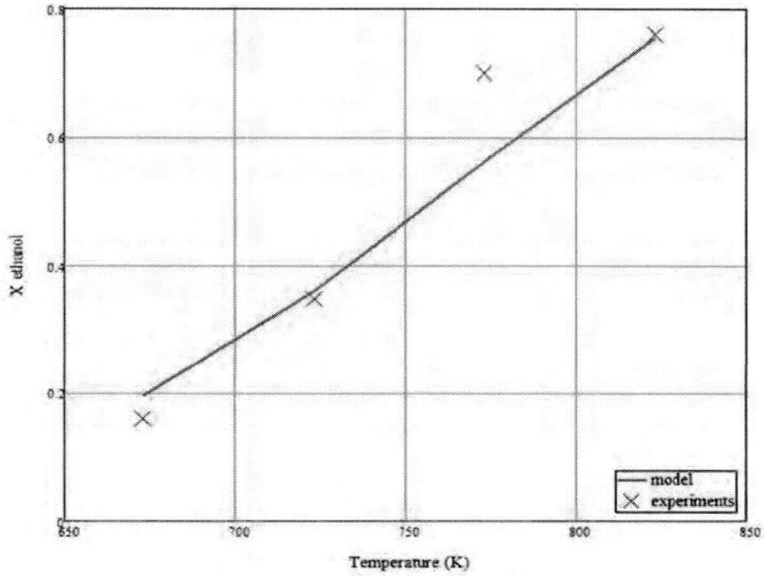

Model Validation

Department of Chemical & Biomolecular Engineering

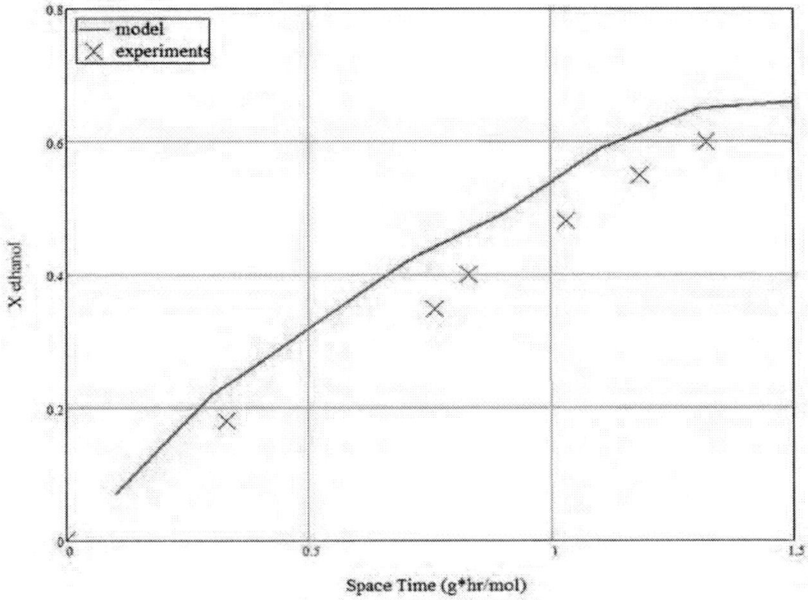

Effect of Hydrogen Flow Rate on Power Output

Department of Chemical & Biomolecular Engineering

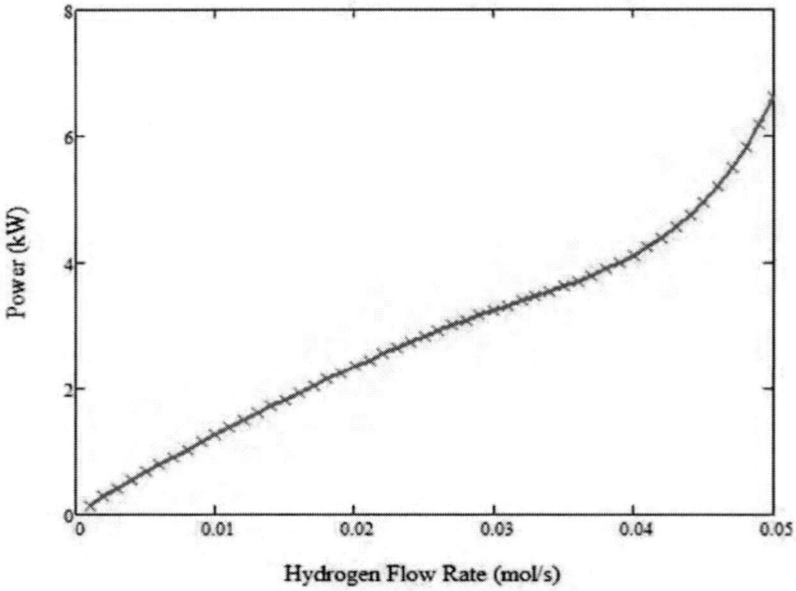

Simulation

Department of Chemical & Biomolecular Engineering

Table 4: Values of reaction parameters for reformer model analysis.

Name	Symbol	Value	Unit
Length of Reformer	L	3	m
Inner Diameter of Reformer	D_i	20	cm
Outer Diameter of Reformer	D_o	20.5	cm
Ethanol Feed Flow Rate	F_e	0.0035	mol/s
Inlet Temperature	T	900	K
Inlet Pressure	P	1	atm
Void Fraction	ϕ	0.3	
Steam to Ethanol Ratio	r_s	3	
Diameter of Particle	D_p	0.3	mm

Temperature profile

Department of Chemical &
Biomolecular Engineering

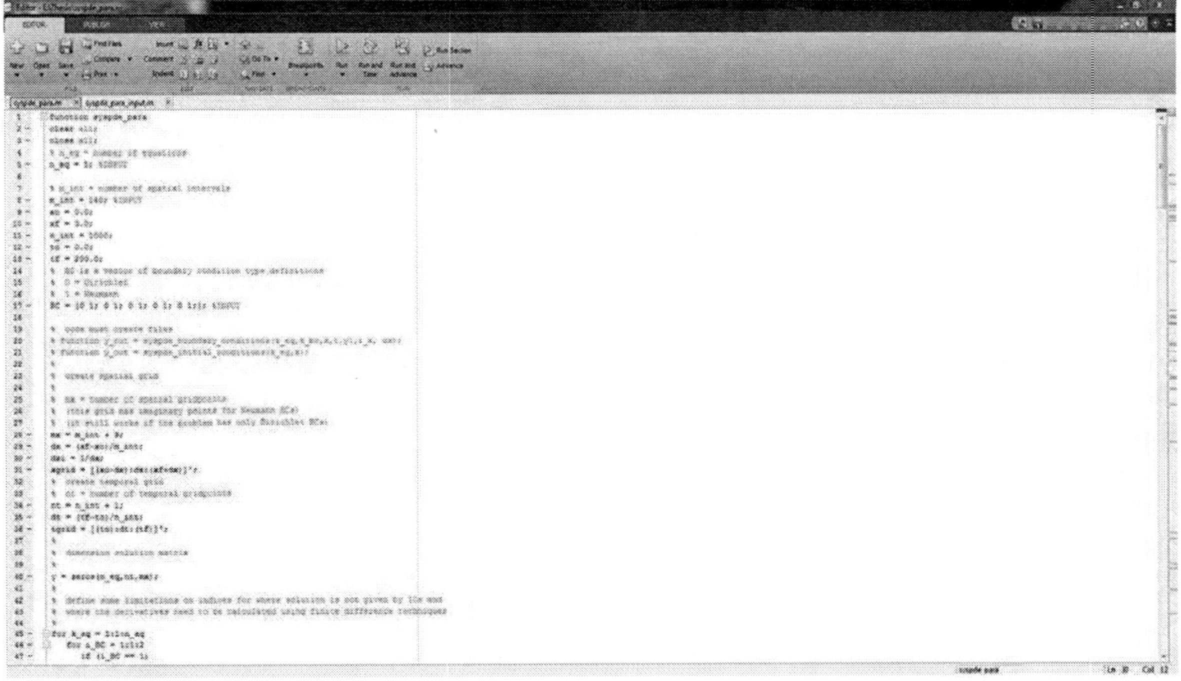

Concentrations Profile

Department of Chemical &
Biomolecular Engineering

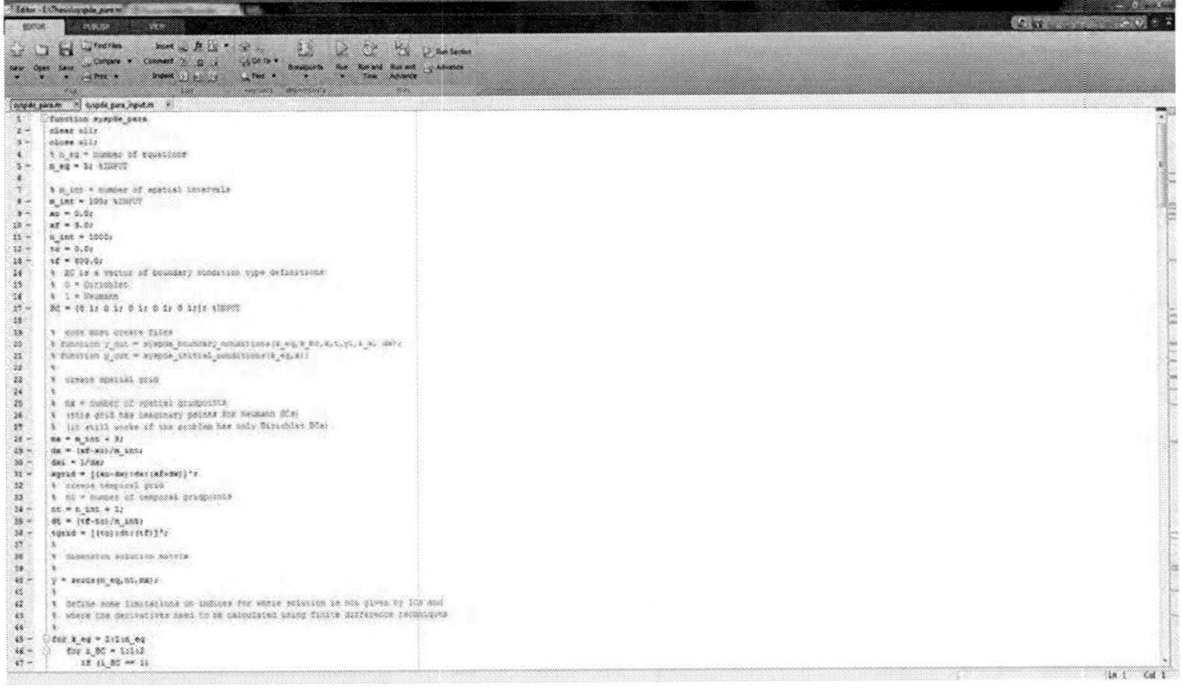

Effect of Inlet Temperature on the Conversion of Ethanol

Department of Chemical & Biomolecular Engineering

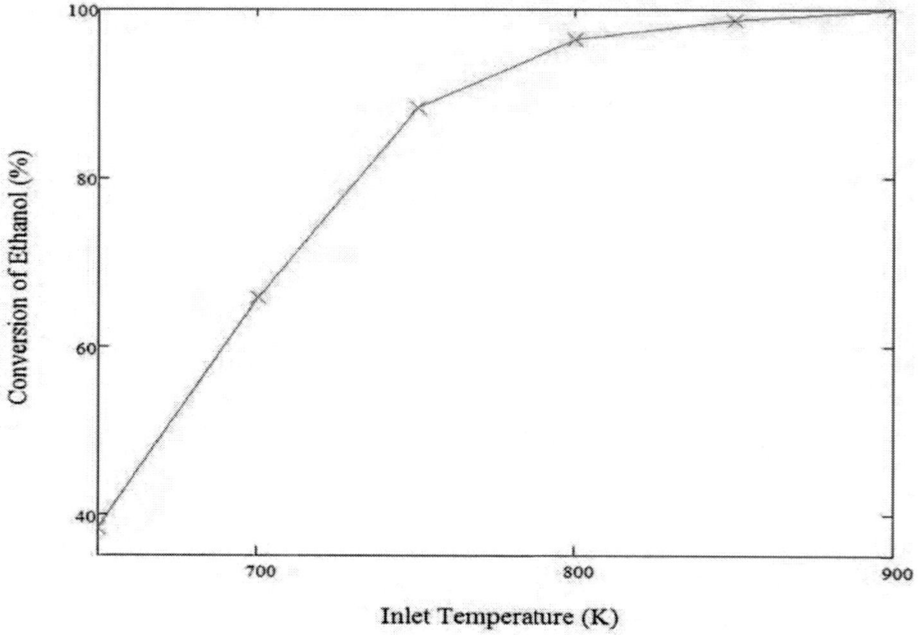

Effect of Reformer Length and Diameter on The Conversion of Ethanol

Department of Chemical & Biomolecular Engineering

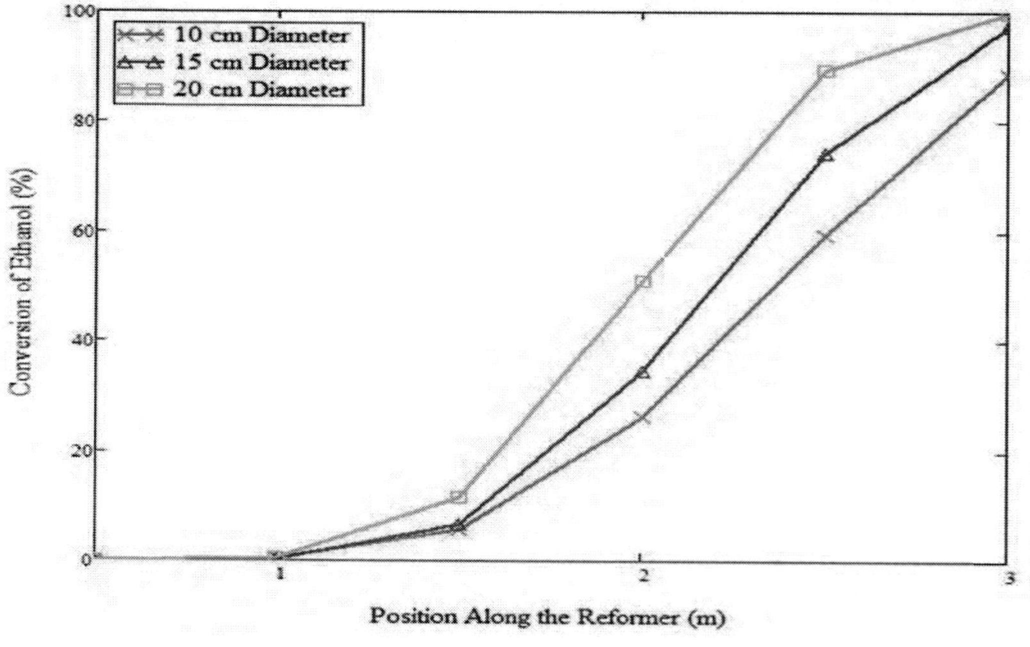

Effect of Ethanol Feed Flow Rate on The Conversion of Ethanol

Department of Chemical & Biomolecular Engineering

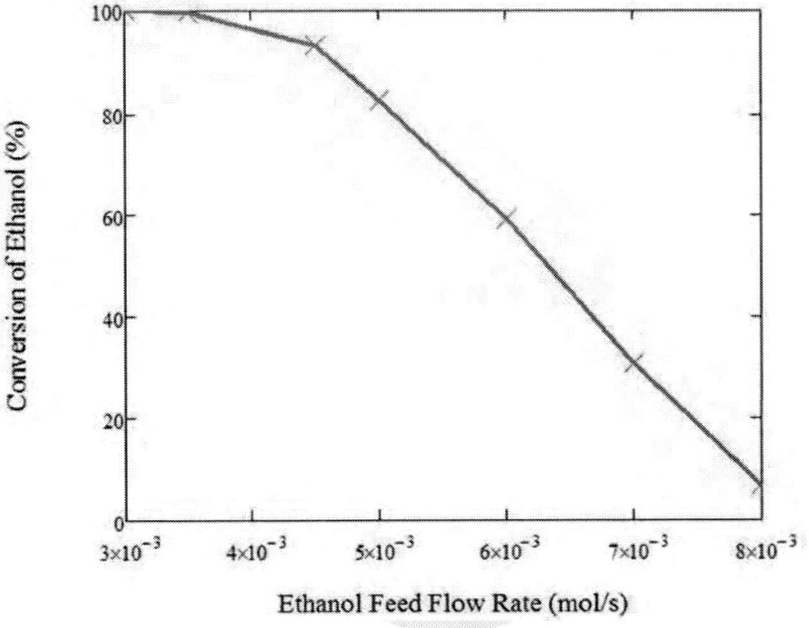

Effect of Ethanol Feed Flow Rate on The Hydrogen Yield

Department of Chemical & Biomolecular Engineering

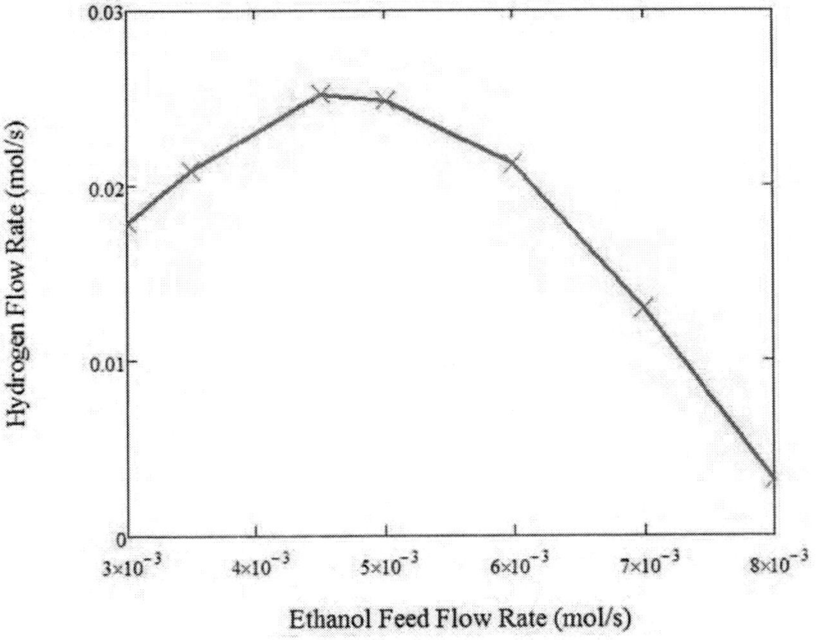

Effect of Time on the Conversion

Department of Chemical & Biomolecular Engineering

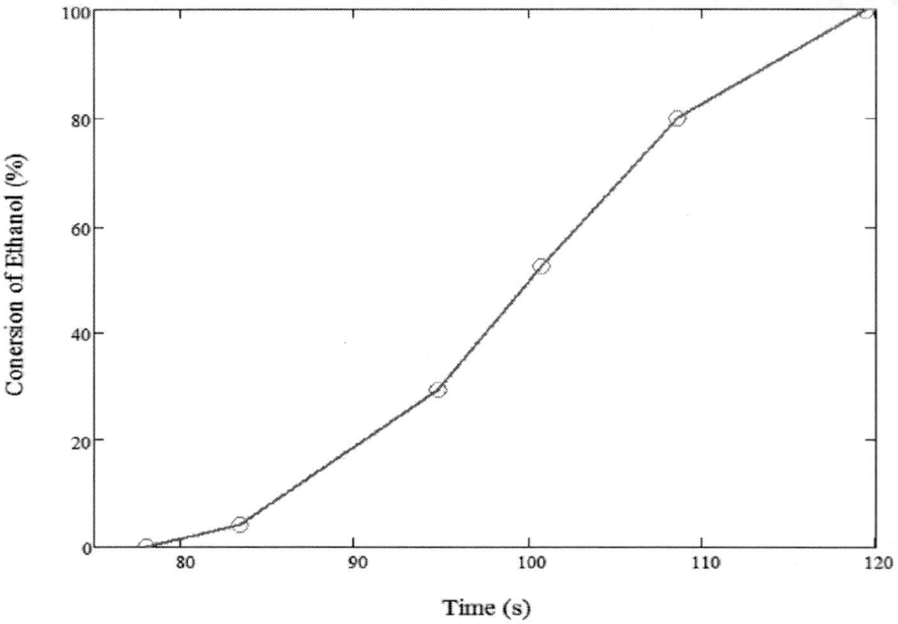

Effect of Particle Diameter on the Dispersion Coefficient

Department of Chemical & Biomolecular Engineering

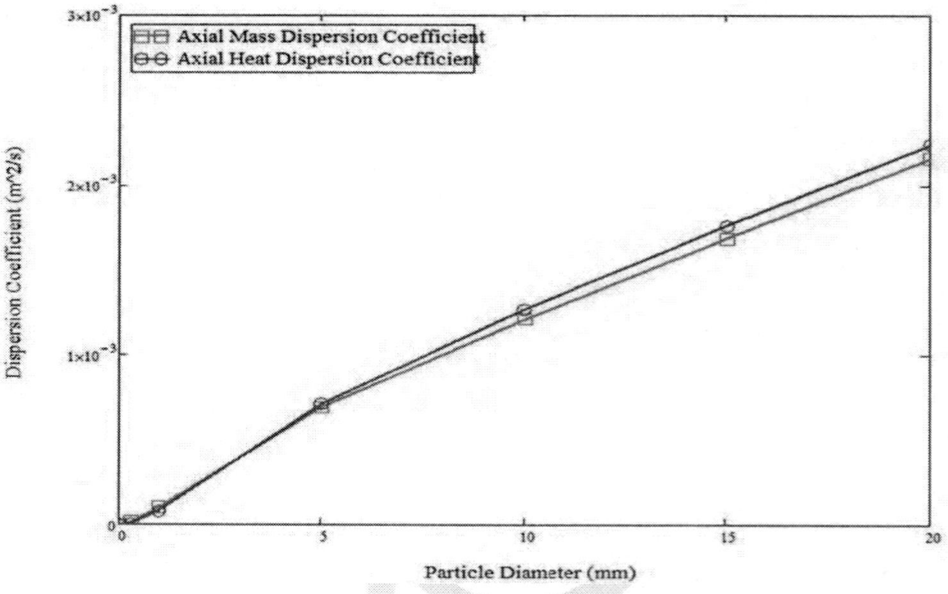

Effect of Particle Diameter on the Conversion of Ethanol

Department of Chemical & Biomolecular Engineering

Table 4: Values of effectiveness factors for different particle diameters where the non-isothermal particle effect is considered.

Particle Diameter (mm)	Ethanol Conversion (%)
0.003	100
0.03	100
0.3	100
1	100
5	100
10	95.2
15	81
20	57.1

Effect of Inlet Pressure on the Conversion

Department of Chemical & Biomolecular Engineering

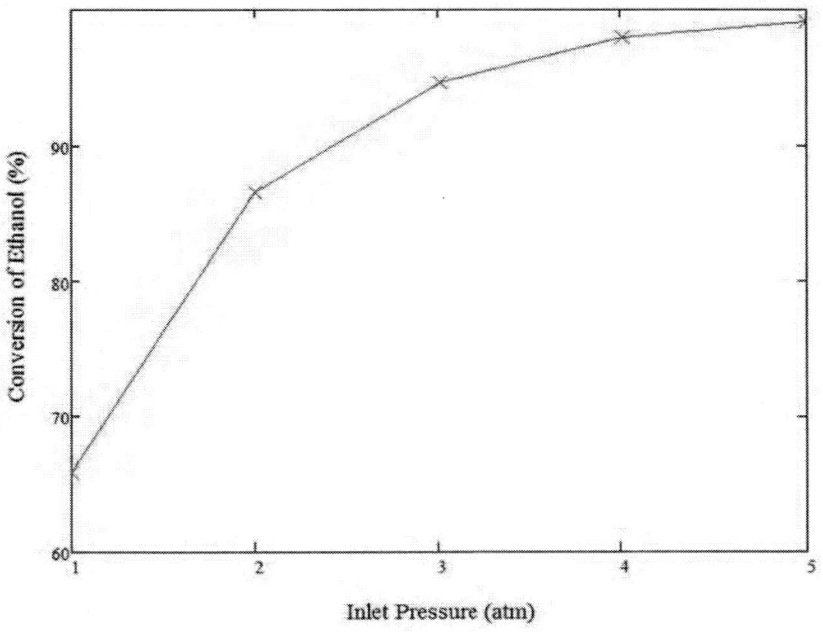

Effect of Inlet Pressure on the Jacket

Department of Chemical & Biomolecular Engineering

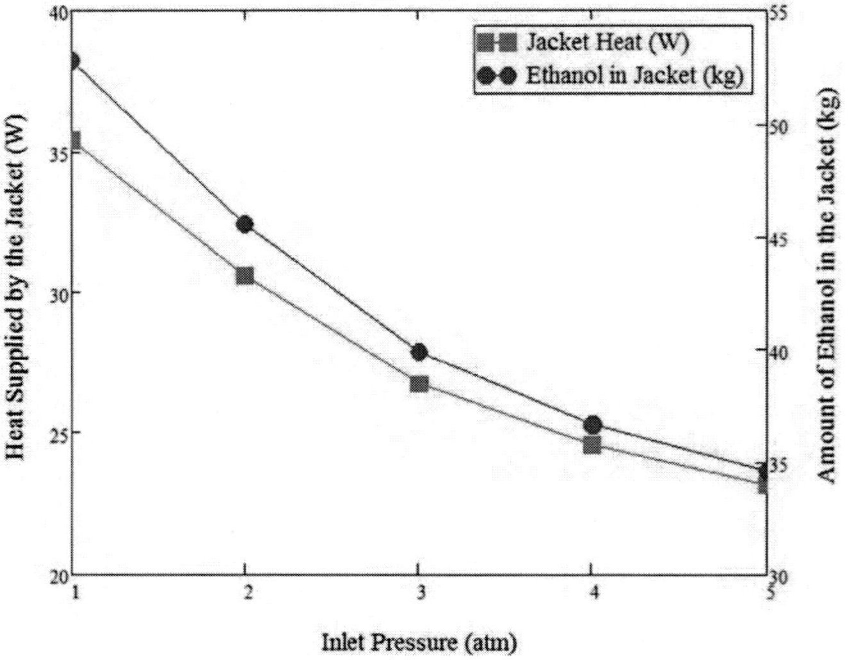

Reformer Model

Department of Chemical & Biomolecular Engineering

Table 5: Values of reaction parameters for reformer model.

Name	Symbol	Value	Unit
Length of Reformer	L	4	m
Inner Diameter of Reformer	D_i	30	cm
Outer Diameter of Reformer	D_o	30.5	cm
Inlet Temperature	T	1000	K
Inlet Pressure	P	1	atm
Steam to Ethanol Ratio	r_s	3	
Diameter of Particle	D_p	0.3	mm

Power Supplied Vs. Power Demands

Department of Chemical & Biomolecular Engineering

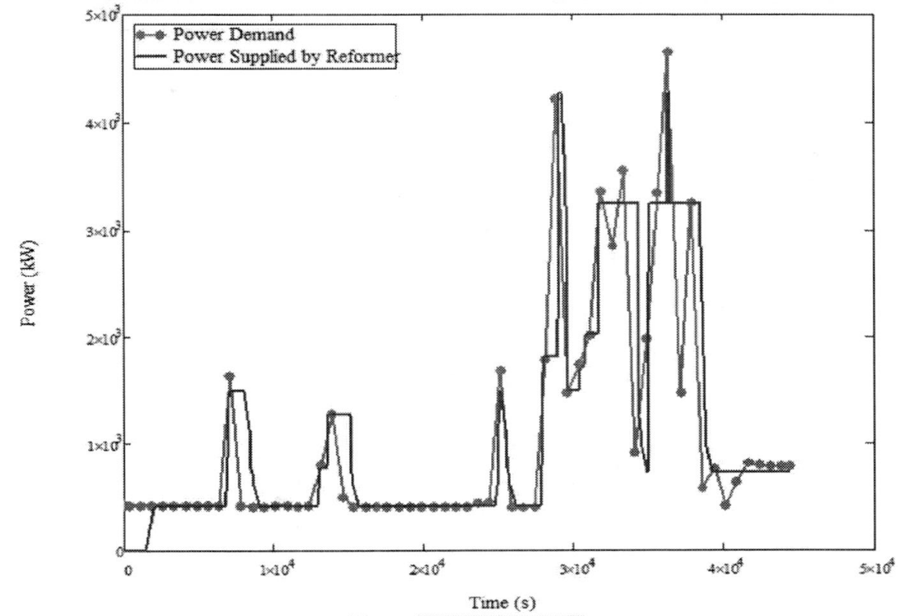

Power Lag Profile

Department of Chemical & Biomolecular Engineering

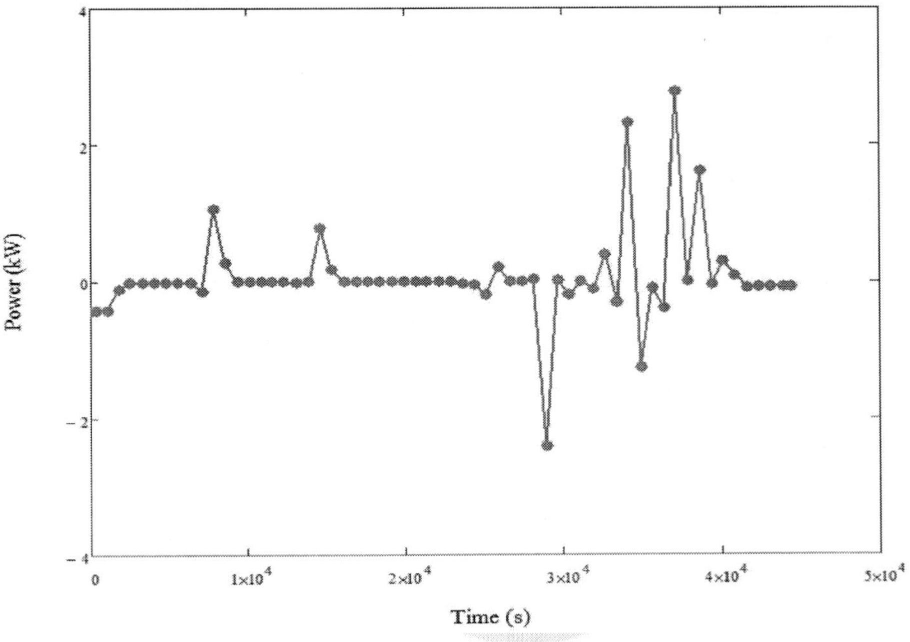

Back-up Battery Profile

Department of Chemical & Biomolecular Engineering

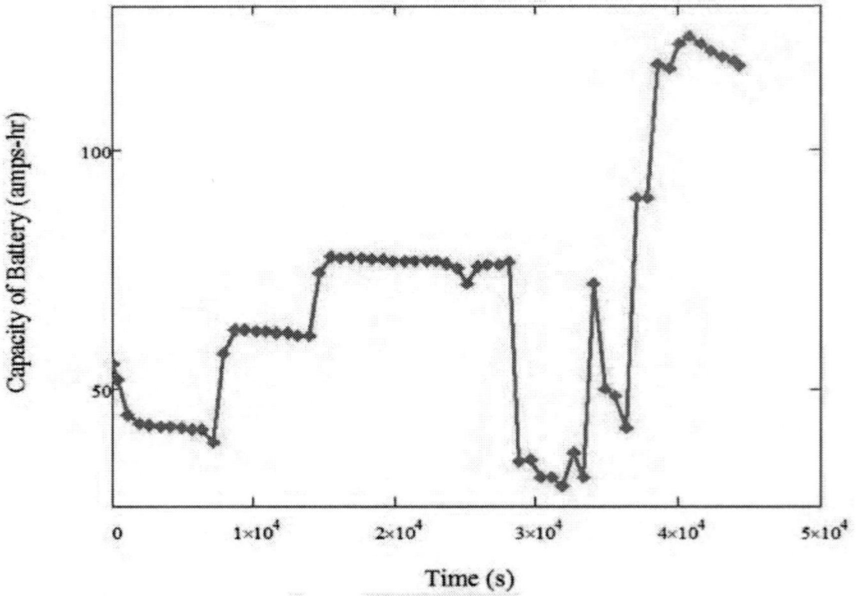

Reformer Evaluation

Department of Chemical & Biomolecular Engineering

Table 6: Reformer evaluation.

Parameter	Unit	Value
Length of Reformer	m	4
Area of Reformer	m^2	3.833
Inlet Temperature	K	1000
Inlet Pressure	atm	1
Power Needed	kW	-2.9
Excess Power Stored	kW	4.7
Ethanol Jacket Feed	kg/cycle	63.48
Ethanol Feed	kg/cycle	3.285
Voltage of Battery	V	12
Battery Material		Rechargeable Lead-Acid
Capacity of Battery	amps-hr	55

Results

Department of Chemical & Biomolecular Engineering

Table 7: Efficiency and cost comparison of ethanol to diesel and natural gas.

Parameter	Unit	Ethanol	Diesel	Natural Gas
Efficiency	%	64.1	32.3	26.8
Fuel Cost	$/kg	1.055	0.937	0.498
Cycle cost	$/cycle	74.1	64.5	34.3

Significance of the Work

Department of Chemical & Biomolecular Engineering

- It provides the first mathematical model to estimate the dynamic behavior of a steam reformer.
- It Provides the first steam reformer design for a residential application.
- It provides a matrix of research topics for future students.
- It is a good option for an energy self-sufficient house, farm, or building which comes from pure renewable energy source.
- It provides a good numerical solution package for solving unsteady-state and non-isothermal multi-reactions reactor.

Conclusion

Department of Chemical & Biomolecular Engineering

- Fuel Cells provide a good alternative as a source of energy
- The model gives a good prediction to the experimental data.
- The model gives a good prediction to wide varieties of reaction parameters.
- The conversion of ethanol increases as the inlet temperature, diameter, pressure, time, and length of the reformer increases.
- Reformer can supply the power demand with an excess power that can be stored. However, a back-up battery is still needed for start-up.

The End

Department of Chemical & Biomolecular Engineering

Questions ????

390063 Synthesis and Transportation Characteristic of Poly (aryl ether oxadiazole) Amphoteric Ion Exchange Membrane for Redox Flow Batteries: Experimental and Molecular Dynamics Simulation

Abstract:

All vanadium redox flow battery (VFB), a novel large-scale energy storage device, has drawnmore and more attention in recent years, and be considered as the potential technology in new energy generations application. Ion exchange membrane for redox flow battery applications provides conductive channel and prevents the crossover of different vanadium ions between positive and negative electrolytes. Currently the main membrane employed in VFBs is perfluorosulfonic acid cation exchange membrane (Nafion® series), it owns high ion conductivity and stability, however suffers from poor ionic selectivity. The positive exchange groups can prevent vanadium ions crossover because of the Donnan effect. However, the ionic conductivity of anion exchange membrane (AEM) is much less than that of cation exchange membrane (CEM), resulting the poor charging and discharging performance of VFB. Amphoteric ion exchange membrane (AIEM), with both cation and anion exchange groups in the membrane matrix, is endowed with excellent properties in conductivity and crossover-resistance.

In this study, a new route for synthesis of a novel amphoteric ion exchange membrane has beendeveloped. Compounds with amino-group and sulfonic group is employed as functionalized reagent, and poly (aryl ether oxadiazoles) as basic copolymer, a series of poly (aryl ether oxadiazole) amphoteric ion exchange membrane (AIEM-x) are synthesized and tested by GPC, NMR, FTIR, vanadium permeability and ion conductivity. Molecular dynamics simulation of AIEM-x is analyzed based on the first-principle. The experimental and simulation results indicate that the 2-aminoethanesulfonic acid branch structure in AIEM-x has both cation and anion group, and the groups can attract each other and become an ion cluster when membrane-forming. Electron cloud density of the ion cluster part in the membrane grows significantly, so the ion clusters turn into ionic transportation channels and reduce the resistance of the ions transport though the membrane, so the ion conductivity improves. Because of using of 2-aminoethanesulfonic acid, the functionalization of cation and anion exchange groups is one-step reaction and in mild condition, besides it is low in cost and high in productivity. The poly (aryl ether oxadiazole) amphoteric ion exchange membrane shows a good characteristics in ion conductivity, mechanical and heat stability, acid-resistance, is a promising membrane applied in the redox flow batteries.

Copyright © American Institute of Chemical Engineers. All rights reserved.

392629 Understanding the Gas Diffusion Layer in PEM Fuel Cells: Influence of the PTFE Distribution on Ex-Situ Properties and in-Situ Performance

Abstract:

Proton exchange membrane fuel cells (PEMFCs) are being explored for use in transportation, as well as back-up and portable power applications. A number of key technical and cost challenges remain before they can be used on the commercial scale, however. The Gas Diffusion Layer (GDL) is one of the most critical components for PEMFC and it is of paramount importance to understand how key properties influence performance and ultimately cost. This paper focuses on the PTFE distribution in the GDL, but more specifically on how to control the PTFE distribution throughout the GDL, and the impact of the distribution of PTFE in the GDL on key *ex-situ* properties including porosity, pore size, thickness under compression, resistivity under compression, diffusivity, through-plane permeability and in-plane permeability. The influence of PTFE distribution on *in-situ* performance will also be considered under both normal and extreme operating conditions.

Copyright © American Institute of Chemical Engineers. All rights reserved.

392440 Electrocatalytic Processing of Renewable Biomass-Compounds to Generation of Electricity, Chemicals and Fuels

Abstract:

Biomass is abundant, inexpensive and renewable, therefore, it is expected to play a significant role in our future energy landscape. The U.S. DOE has identified some platform compounds (or building blocks, i.e. glycerol, levulinlic acid, etc.) that can be mass-derived from biomass, currently the chief research need is to develop cost-effective, green and sustainable approaches to utilize these biorenewable compounds. I will present our recent efforts on exploration of aqueous phase electrocatalytic processing of the renewable biomass platform compounds to production of electricity, valuable chemicals and biofuels. We discovered that electrocatalytic oxidation of biorenewable polyols to valuable products with a tuneable degree of oxidation can be achieved by varying electrode potential. The degree of glycerol oxidation over Au nanoparticles was tuned with anode potential to produce tartronate (oxidizing two primary -OH groups, ≥ 0.35 V), mesoxalate (oxidizing three -OH groups, ≥ 0.45 V), or glycolate (breaking C-C bond, ≥ 0.9 V), and the mechanism of potential-controlled glycerol electro-oxidation has been proposed. This may open a new strategy for the targeted transformation of biomass compounds with poly or multi- functional groups into valuable chemicals. We further demonstrated high power density anion exchange membrane fuel cell with crude glycerol fuel using carbon supported PtCo dealloyed catalyst. We also explored electrocatalytic processes to store renewable electricity in biofuels, and have demonstrated electrocatalytic reduction of levulinic acid to higher energy-density biofuel intermediates: valeric acid (VA) or γ-valerolactone (gVL) on non-precious Pb electrode with high yield and faradaic efficiency in a single electrolysis flow cell reactor. The applied potential and electrolyte pH were found to control the reduction products. It is interesting that formic acid (coproduced in cellulose hydrolysis stream) can improve the rate of electro-reduction of LA to VA. Our work shows promise for direct electrocatalytic processing of crude biorefinery streams to electrical energy and biofuels/intermediates production.

Copyright © American Institute of Chemical Engineers. All rights reserved.

386209 PEMFC System Using Metallic Hydride Cycles to Purify the Hydrogen

Abstract:

Energy production trough hydrogen utilization has gained attention in the last decades. Reformer reactors are used to produce hydrogen rich streams departing from hydrocarbons and alcohols. The specifications of hydrogen needed depend on the device used to produce energy. Hydrogen containing low CO concentrations is required for proton exchange membrane fuel cells (PEMFC). Different purification processes were considered to achieve this aim. This work presents a system involving a reforming process and a PEMFC device coupled to pressure swing absorption (PSAb) purification reactor. Then, the PEMFC system has the following components: reformer reactor, PSAb reactor, PEMFC, heat exchangers, auxiliary units and post combustion system.

The study case used in this work considers ethanol as a fuel. A steam reformer reactor is adopted. So, the inputs at reformer reactor are water and ethanol in an operative molar relation to be optimized. The energy necessary in the reforming reactor is provided by a post combustion system. The output stream in this reactor contains water, carbon oxides and hydrogen. This stream feeds the PSAb reactor. This reactor has solid hydride material that absorbs the hydrogen and allows the impurities separation via venting. Here two PSAb reactors working in parallel are proposed to give a constant flux of pure hydrogen to the PEMFC. A cyclic scheduling of PSAb reactors is presented. The three steps involved in each reactor are: absorption, venting and desorption. The venting impurities are burn in the post combustion system. The PEMFC device produces the usable energy of the system and exhaust gases. These gases are also used by the post combustion system. This system is used to close the energy balance of the entire system.

A mathematical model of energy production system via PEMFC is developed in GAMS. The model calculates and optimizes the efficiency of the system. This allows gaining insight about the optimal values of the operative variables in each device of the system.

Copyright © American Institute of Chemical Engineers. All rights reserved.

Hydrochar derived high-mesopore area carbon as a cathode for Li-ion hybrid electrochemical capacitors

Akshay Jain[a], *Sundaramurthy Jayaraman*[a], *Rajasekhar Balasubramanian*[b] & *M.P. Srinivasan*[*a]

[a] Department of Chemical and Biomolecular Engineering, National University of Singapore, Singapore 117585, Singapore.

[b] Department of Civil and Environmental Engineering, National University of Singapore, Singapore 117585, Singapore

Mesoporous carbons with high surface area were synthesized by using hydrothermal treatment of biomass (coconut shell) in the presence of $ZnCl_2$ as the activating agent. Mesopore development is shown to be promoted by conditions that favour increased diffusion of $ZnCl_2$ into the biomass and leads to effective dehydration compared to conventionally employed conditions. Upon physicochemical activation, the hydrochar yielded mesoporous carbon with enhancement of mesopore area by up to 67%.

The mesoporous carbons were used as a cathode material in a non-aqueous medium for the fabrication of the Li-ion Hybrid Electrochemical Capacitors (Li-HEC) with commercially available $Li_4Ti_5O_{12}$ serving as the anode. Increase in mesoporosity by the use of hydrothermal treatment was manifested as a significant energy density of ~ 69 Wh kg^{-1} relative to that obtained from commercial activated carbons (36 Wh kg^{-1}) and excellent cycleability ~ 2000 cycles of the Li-HEC. The study underscores the utility of simple pre-treatment protocols that increase the efficiency and effectiveness of the adsorbent matrix for development of high energy density Li-HEC.

Viscosity and CO_2 Diffusivity of Liquid-like Nanoparticle Organic Hybrid Materials determined using ATR FT-IR Spectroscopy

Kun-Yi Andrew Lin[1], *Camille Petit*[2] *and Ah-Hyung Alissa Park*[3]

[1]Department of Environmental Engineering, National Chung Hsing University, 402 Taiwan R.O.C.

[2]Department of Chemical Engineering, Imperial College London, SW7 2AZ, UK

[3]Departments of Earth and Environmental Engineering and Chemical Engineering and Lenfest Center for Sustainable Energy, Columbia University, New York, NY 10027

ABSTRACT

Novel liquid-like hybrid nanomaterials, referred to as Nanoparticle Organic Hybrid Materials (NOHMs), were synthesized by grafting polymer chains onto silica nanocores. This new class of materials has shown promises for a number of applications, in particular CO_2 capture. Owing to their enhanced thermal stability, negligible vapor pressure and chemical and physical tunable nature, NOHMs represent a potential alternative CO_2 capture solvent to the commonly used amine-based solutions. Aside from the absorption mechanisms, there are a number of aspects to investigate when considering NOHMs for the CO_2 removal technology, as for instance CO_2 diffusivity and viscosity of the materials, which are both linked to the absorption kinetics. This study was, therefore, conducted to investigate the CO_2 diffusivity and viscosity of different NOHMs as a function of NOHMs' structural parameters and temperature. Towards this goal, NOHMs with different structures (i.e., different core fraction and core size) were synthesized, characterized and evaluated via an FT-IR spectroscopic approach, which allowed synchronized measurements of CO_2 capture capacity and diffusivity. The results indicate that the core size has a negligible impact on the viscosity and CO_2 diffusivity of NOHMs, while lower core fractions and higher temperatures significantly enhanced the kinetics of CO_2 absorption in NOHMs. Furthermore, a first attempt to model the CO_2 diffusion phenomena in different NOHMs was carried out and the predicted trends were in good agreement with the experimental observations.

360686 Dynamic Wetting Effect on 3-D Liquid Film Waving for CO2 Absorption Intensification

Abstract:

CO_2 capture has been increasing demand for global energy saving and environment protection. It is noticed that this chemisorption process in the two-phase contact system interact closely with flow behavior. Given most literatures focused on absorbent capability or Marangoni convection on gas-liquid interface, this work investigated solid substrate property and its effect on flow behavior and finally influenced the removal rate. To improve the chemisorption efficiency, we fabricated metal wire mesh as solid substrate on which absorbent solution film formed. MEA (monoethanolamine) with concentration of 15wt% was used as aqueous absorbent, and experiment showed that such solutions were well wetted on the surface of wire mesh substrate. To make comparison, smooth flat plate was adopted in the experiment and it manifested that on the flat plate, liquid drop showed clear triple-phase contact line between free surface and solid substrate. It was also illustrated by OCA15 facility the difference of contact line of liquids in static or slippage state.

Both experiment and numerical simulation indicated that dynamic wetting had great potential to change flow behavior. The results manifested that in good wetting case, liquid film was well spread; however in worse wetting, stagnation point was found in the middle of the flow region. Due to the flow behavior change, CO_2 capture rate can be increased by 2~3 times. Thus this fabricated wire mesh can be used as a novel structured packing for such chemisorption processes.

In the simulation model, three-dimensional liquid film structure is reported for the first time, and also combined with its chemisorption behavior. Drag force and surface tension were introduced into momentum equations as source terms in the computational fluid dynamic model. In this thin film contact process, interfacial turbulence was acquired its effect on component diffusion intensification. Dynamic wetting property was also considered in the flow model. 3-D liquid film images were illustrated to show how this fabricated wire mesh could intensify CO_2 capture.

Copyright © American Institute of Chemical Engineers. All rights reserved.

CO_2 -Plume Geothermal (CPG) system performance for various operating and geologic parameters.

Nagasree Garapati[1], Jimmy B. Randolph[1], and Martin O. Saar[1]

[1]Department of Earth Sciences, University of Minnesota, Minneapolis, MN.

CO_2-Plume Geothermal (CPG) energy capture involves injection of CO_2 as the working fluid into naturally permeable sedimentary basins for geothermal energy extraction. The injected CO_2 forms a large subsurface CO_2 plume that extracts heat energy from the reservoir, whereafter the geothermally heated CO_2 is produced to the surface to drive a power system, is cooled and reinjected into the subsurface. While the heat density of sedimentary basins is typically relatively low, the large volume of such natural reservoirs, compared to conventional and hydrofractured reservoirs, may counteract this limitation and in some cases likely surpass the total heat energy output possible from the latter. Furthermore, supercritical CO_2 has a high mobility (i.e., low kinematic viscosity) and a high thermal expansibility compared to water, resulting in the formation of a strong thermosiphon that typically eliminates parasitic pumping power requirements. Both effects significantly increase electricity production efficiency. Because the produced CO_2 is reinjected into the deep saline aquifers along with the main CO_2 sequestration stream coming from a CO_2 emitter, CPG systems can be operated as CO_2 sequestering geothermal power plants that have a negative carbon footprint.

In this work we model the CO_2 plume development in the reservoir and the geothermal energy extraction from the reservoir over time using the TOUGH2 numerical simulator with the equation of state module ECO2N. The current model is three-dimensional but axisymmetric, allowing significant reductions in computational effort. The cylindrical coordinate system simulates symmetrical CO_2 plume formation around the vertical, over the reservoir depth fully penetrating, injection well. The model further consists of a 50 m thick reservoir that is located at average depths from 1.5 km to 5.0 km. The radial dimensions to the production well vary from 0.7 km to 2.0 km, while the numerical model extends to a radial distance of 100 km to minimize boundary effects. The production well is horizontal, circular, and located at the top of the reservoir at a distance from the injection well, which allows the CO_2 extraction operations to take advantage of buoyant CO_2 flow over the brine and maximize CO_2 recovery while minimizing unwanted brine extraction. Although it constitutes a simplified geometry, the radial configuration accurately captures physical CO_2/brine multifluid behavior, including modern fluid production from a horizontal production well, while permitting efficient numerical modeling.

Here, we present results for the performance of a CPG system as a function of various operating and geologic parameters such as distance between vertical injection and horizontal production wells, CO_2 circulation rates, geothermal gradient, reservoir depth, and vertical permeability heterogeneity of the layered reservoir. The heterogeneity of the reservoir is studied by dividing the reservoir into multiple horizontal geologic layers, each layer with a different horizontal permeability, k_x. The vertical permeability, k_z, is adjusted to maintain the same anisotropy ratio of $k_x/k_z=2$ in all scenarios. We determine the amount of CO_2 required for operation of a geologically simplified CO_2-Plume Geothermal (CPG) system as well as reservoir heat depletion and lifetime, and fluid pressure build-up during operation as a function of various parameters such as reservoir volume, temperature, and depth. We find that the amount of CO_2 required for extended CPG operation increases with reservoir thickness, well spacing between injection and production wells, and reservoir depth but decreases for higher geothermal gradients. While the rate of produced fluid temperature depletion decreases with reservoir thickness and well spacing between injection and production wells, the time taken for the temperature to start depleting is less for higher geothermal gradients and at shallow and higher depths. We further find that during the build-up of fluid circulation to an identical maximum fluid circulation rate, the pressure profile and the CO_2 fraction in the produced fluid varies with the permeability and the thickness of the heterogeneous layers in the reservoir.

Iron-Containing Mixed-Oxides for Chemical Looping with Oxygen Uncoupling (CLOU)

*Arya Shafiefarhood, Amy Stewart, Fanxing Li**

Department of Chemical and Bimolecular Engineering

North Carolina State University

Raleigh, NC 27695

* Email: fli5@ncsu.edu

Chemical looping with oxygen uncoupling (CLOU) offers a potentially effective approach for solid carbonaceous fuel conversion. The CLOU process is based on oxygen carriers that can reversibly exchange their lattice oxygen with external environment. Spontaneous oxygen release at low oxygen partial pressures, ease of regeneration at high oxygen partial pressures, and high accessible oxygen content are required properties for oxygen carriers in the CLOU process. The varying and often tunable thermodynamic properties of pure and mixed oxides of first row transition metals make them potentially viable candidates for CLOU applications. In this study, mixed oxides of iron-cobalt and iron-manganese are synthesized and tested for their CLOU properties. The effect of addition of secondary perovskite phase, as support, on uncoupling properties of these oxides is also investigated. Experimental results indicate that different combination of cations results in oxides with different thermodynamic properties. Oxides with higher concentrations of Co and Mn generally showed lower decomposition temperatures. Presence of perovskite support is found to notably affect the uncoupling properties of Fe-Co and Fe-Mn mixed oxides.

388202 Feasibility of Integration of Chemical Looping Combustion (CLC) in IGCC Systems: High Pressure Clc Experiments with CH4 and Syngas

Abstract:

The effect of elevated pressure on chemical looping combustion using nickel and copper oxygen carriers was studied in order to evaluate the feasibility of integration of CLC into IGCC systems. Increasing the operating pressure of an IGCC + CLC system increases the thermal efficiency of the process, while decreasing the compression energy demand for high pressure CO_2 recovery.[1] Methane and syngas combustion experiments were conducted in a high-pressure fixed bed reactor, sketched in Figure 1, at system pressures up to 10 bar. Consistent data were achieved over multiple cycles of chemical looping for both oxygen carriers, indicating successive cycle regenerablity. The effect of pressure on the oxidation and reduction kinetics was studied for both oxygen carriers in order to propose reaction kinetics, based on the experimental data. Characterization of the oxygen carriers after successive cycles was performed via SEM, XRD, BET and XPS, indicating significant structural changes as the result of the high pressure. Figures 2 and 3 show experimental results of the reduction step of both oxygen carriers at different system pressures and temperature of 800 °C. A significant increase in CO_2 selectivity is observed at the high-pressure experiments, which is especially clear for the experiments with Cu oxygen carrier. In the case of the Ni-based oxygen carrier carbon formation during reduction is reduced indicating a significant change in the carbon formation and consumption kinetics. These observations were analyzed using a fixed-bed reactor model indicating promising applicability of CLC for industrial CO_2 separation.

Table 1: Experimental Conditions

Oxygen Carrier	20% NiO/Al_2O_3, 36% CuO/SiO_2
Load	2.2 grams
Reactor Dimensions	I.D. 9.6 mm, height 482 mm
Total Flow	100 sccm
Temperature	1073 K
Pressure	1, 5, 7.5, 10 bar
Oxidation	8 min, 10% O_2/Argon
Reduction (CH_4)	2-3 min, 10% CH_4/Argon
Reduction (syngas)	3 min, 5% CO, 5% H_2 /Argon

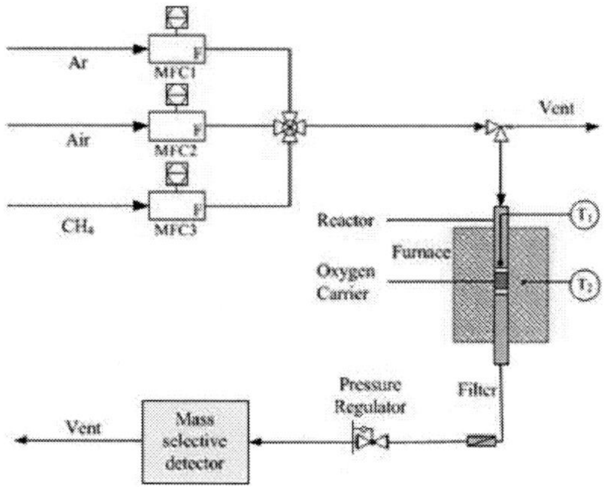

Figure 1: Fixed-bed chemical-looping setup used in this work.

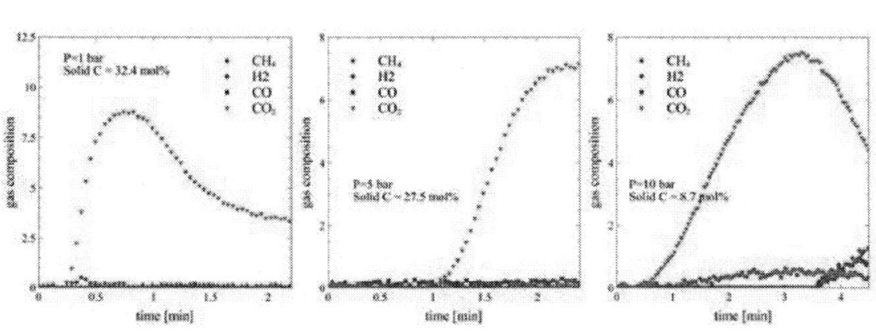

Figure 2: Reduction of NiO/Al_2O_3 and CH_4/Ar in a fixed-bed reactor at 800 °C at 1, 5 and 10 bar including the solid carbon deposition.

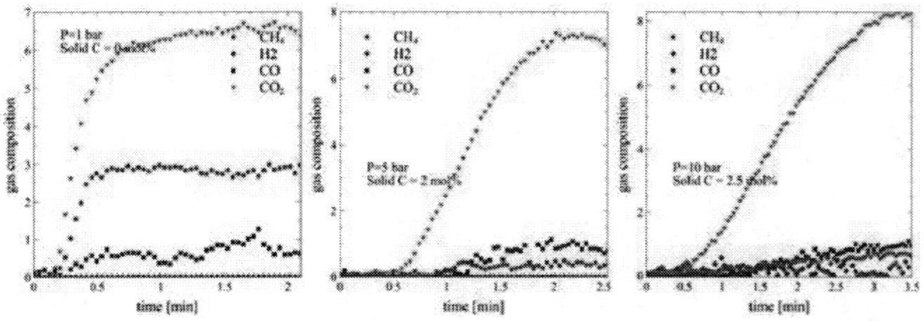

Figure 3: Reduction of NiO/Al_2O_3 and CH_4/Ar in a fixed-bed reactor at 800 °C at 1, 5, and 10 bar .

Acknowledgement: This material is based upon work supported by the National Science Foundation under Grant No. 1054718.

References:

[1] *IPCC Special Report on Carbon Dioxide Capture and Storage*; Cambridge University Press: New York, 2005; available on the web athttp://www.ipcc.ch.

Granular Modelling of Gas-Solid Hydrodynamics in Fluid Catalytic Cracking (FCC) Riser Reactors using Commercial Code

Engr. Dr. M. N. Idris*; MAIChE, AMIChemE, MNSE
Energy & Process Refining Technology Group
Department of Chemical Engineering
University of Maiduguri
Borno State, Nigeria
Tel: +234 (0) 705 877 1096
Email: idrismn@unimaid.edu.ng, muhinu@hotmail.co.uk

ABSTRACT

The application of commercial computational fluid dynamics (CFD) usefulness to study the hydrodynamics of gas-solid flow in FCC is a complex phenomenon. The essence of this study is basically on the heat transfer conservation and economies. In order to establish concise and reasonable predictions, a choice of runs for time-average comparison for solid velocities and concentrations to measure fluxes and densities along the riser axes of heat transfer were used. The model structure were carried out successfully and compared with experimental data on a CFB/FCC riser length of 15.1m and 0.10m in diameter. The model incorporates a two-fluid framework with kinetic theory of granular flow (KTGF) to simulate fully developed gas-solid flows in vertical risers. This study presents the simulation of gas-solid hydrodynamics and flow behaviour in a riser circulating fluidised bed (CFB) reactor using the Eulerian-Lagrangian two-fluid modelling approach, incorporating a kinetic theory constitutive model for dilute assemblies of FCC particulate solid. The interaction between gas and particles was modelled using particle interphase momentum transfer model with Schiller-Naumann drag model on a proprietary CFD code (ANSYS CFX), using turbulent k-□ models. The predicted heat transfer coefficients were reasonably compared with published experimental data. The overall flow patterns within the fluidise bed riser were predicted well by the model.

The results of this study help in predicting the commercial implications and its economics toward the choice of heat conservation and utilisation in the FCC process systems.

Keywords: *CFD, CFB, KTGF, Heat transfer, Hydrodynamics modelling*

Correspondence concerning this article: *Please for all correspondence on this article, kindly contact the principal author as stated above, or School of Process, Environmental and Material Engineering (SPEME), University of Leeds, LS2 9JT, United Kingdom. Thank you.*

Micro fluidics is gaining immense importance as a separate field of study because of the miniaturization technology. These days the state of the art chemical engineering devices are no longer appears in a giant size, the size of these devices has been reduced to the micro to nano meter ranges. Micro fluidics has marked its presence in a wide area of application such as from lab on cheap devices to micro reactors. Fluid flow in micro channel is a very important aspect of micro fluidics, the detail study of this is needed to fabricate the miniaturized devices.

The study of micro channel flow under the influence of combined pressure and electric field is nothing new, the electro osmosis phenomena was first reported by Reuss in early part of the eighteenth century. Eringen (1964) proved that the micro channel flow characteristics differ from the classical Navier-stokes equation. Tuckermann et al. (1981, 1982) investigated experimentally and found that the flow friction measurements were higher than that predicted by the classical theories. Mala et al. (1997) investigated the interfacial electrokinetic effects on characteristics of liquid flow in a micro channel. Jain et al. (2007) has studied analytically the electrokinetic effects and heat transfer characteristics in a micro channel. Vennela et al. (2011) studied analytically the mass transfer characteristics of neutral solute when a Newtonian fluid is flowing through a rectangular microchannel. Mondal et al. (2013) studied the mass transfer characteristics of neutral solute in a perforated micro tube for a non-Newtonian power law fluid.

Blood by rheology is a non-Newtonian fluid. Many attempts have been done to track the flow characteristics of blood by assuming different rheological model for it. The Casson model is one of the proposed realistic models for blood rheology. In the present work a fluid of Casson model is taken with a symmetric electrolyte and neutral solute dissolved in it and is passing through a perforated micro tube under the influence of combined pressure and electric field.

Detailed study has been done to track the combined velocity profile and expression for the length average Sherwood number for the neutral solute. The permeate side concentration for the neutral is also tracked by following a standard algorithm.

The flow geometry here is a microtube with perforated wall. An electric potential is applied to the wall of the tube which results in development of an electrical double layer. The potential developed at the wall is assumed to be equal to the zeta potential. In the present scenario it is assumed that the zeta potential is less than 25mv. so the Debye-Huckel approximation can be applied to solve the Poisson-Boltzmann equation in order to get the potential distribution in the micro tube. To track the combined velocity profile the Navier-stokes equation is solved with proper assumptions. The electrical body force term is incorporated in the Navier-stokes equation to track the electroosmotic effect in the velocity profile. At first the pressure field is equate to zero to get the expression of the electroosmotic velocity and then electric field is equate to zero to get the expression of the pressure driven velocity. These two velocity profile is summed up to get the expression of the combined velocity profile. To get the expression of the Sherwood number the mass transfer boundary layer equation is solved using the similarity technique. Sherwood number enhancement factor, which gives the enhancement of Sherwood number for the electroosmotic flow over pressure driven flow, is calculated. The tube surface concentration of the neutral solute is calculated using an algorithm. The permeate side concentration is related to the tube surface concentration by real retention. By using the expression of real retention the permeate side concentration profile is also calculated. The flow in the present case happens under the influence of combined pressure and velocity field. These two fields may act in the same direction or may act in the opposite direction. When these two fields act in opposite direction a plane of zero shear appears within the mass transfer boundary layer. The plane of zero shear is the plane on which the solute transport velocity ceases to zero and as a result we see a sudden concentration polarization. This zone is also detected in this study.

References:

Mondal, S, De, S, effects of non-Newtonian power law rheology on mass transport of a neutral solute for electro-osmotic flow in porous microtube, Biomicrofluidics 7, 044113(2013)

Mala, M, Li, Dongqing, Werner, C, Jacobasch, H,J, Ning Y.B, flow characteristics of water through a microchannel between two parallel plates with electrokinetic effects, int. J Heat and fluid flow ,18(1997), 489-496.

Tuckermann, D.B, Pease, R.F.W, optimized convective cooling using micro machined structures, J. Electrochem. Soc.,129, C98

Vennela, N, Mondal, S, Sherwood number in flow through parallel porous plates (microchannel) due to pressure and electroosmotic flow, AIchE, 58(2012), 1693-1703.

Bubble Size and Velocity Measurements in Bubble Columns using Ultrafast X-ray Tomography

Y.M. Lau* and M. Schubert
Institute of Fluid Dynamics, Helmholtz-Zentrum Dresden-Rossendorf,
Bautzner Landstraße 400, 01328 Dresden, Germany
*Corresponding author: y.m.lau@hzdr.de

Highlights

- Dual-plane ultrafast X-ray tomography measurement for bubbly flow is presented.
- Similar bubbles passing the two scanning planes are identified.
- Radial and axial velocities of bubbles are determined.
- Bubble size distribution and gas hold-up are calculated.

Summary

We present measurements of bubble size distribution in bubble columns using X-ray tomography. The experimental setup is an ultrafast electron beam X-ray tomographic system applied to a cylindrical bubble column with a diameter of 0.07 m. Measurements are taken placed on two planes, which are separated with an axial distance of 10 mm. The obtained reconstructed images are filtered, segmented and stacked in the time-domain to form a three-dimensional matrix of bubble objects. By cross-examining the matrices of both planes, we can identify the corresponding bubble objects and determine bubble velocities and volumes as well.

Keywords

Bubble size, Bubble velocity, Bubble columns, X-ray tomography.

Introduction

Bubble columns are widely applied in industrial processes for production of numerous chemicals such as petrol, cosmetics and other daily life products. For the design and scale-up of the bubble column reactors, detailed understanding of the multiphase fluid dynamics behavior and its impact on the heat and mass transfer performance are essential. To study the bubbly flows in bubble columns, the interfacial area is a fundamental parameter, which is usually derived from the bubble size distribution (BSD). The BSD can be measured through both intrusive and non-intrusive experimental techniques. The intrusive measurements, for example wire-mesh sensor [1], are known to disadvantageously disturb the flow by the insertion of the probes. In contrary, this disturbance can be avoided using non-intrusive measurements. One of the commonly used non-intrusive techniques is the digital image analysis (DIA). However, DIA is limited by the visual accessibility and cannot efficiently detect/recognize bubbles at high void fractions, which are present in many bubble columns. There is another promising non-intrusive technique, X-ray tomography [2], which overcomes the DIA limitations mentioned above. Traditionally, analysis of X-ray tomography measurement for a bubble column follows this way [3]: the approximate sizes of the bubbles are determined using the averaged gas velocity via interpolation of the detected object areas in tomographic images of a single plane. In this work, an advanced X-ray tomography analysis technique is developed, which can provide more accurate data of the bubble sizes: actual bubble velocities are determined by identifying the bubbles through tomographic measurement on dual planes; these actual bubbly velocities are applied to calculate the bubble size distribution, by extrusion of the object areas.

Experimental setup

The experimental setup is a cylindrical column with an inner diameter of 0.07 m and a height of 1.50 m. The column is initially filled with liquid (water) of 1.00 m height. Gas (air) is dispersed into the column through a perforated plate, which consists of 95 holes with a diameter of 0.001 m. All the experiments are performed at room temperature and ambient pressure.
In this study, we considered different superficial gas velocities varying from 0.02 m/s to 0.05 m/s, and several measured plane heights, which are 0.30 m, 0.60 m and 0.95 m above the gas distributor. The X-ray measurements are carried out with a frequency of 1 kHz.

Tomographic image analysis

The images are reconstructed from the obtained tomographic data using the filtered back-projection [4]. Examples of reconstructed cross-sectional images of the bubble column are illustrated in Figure 1.

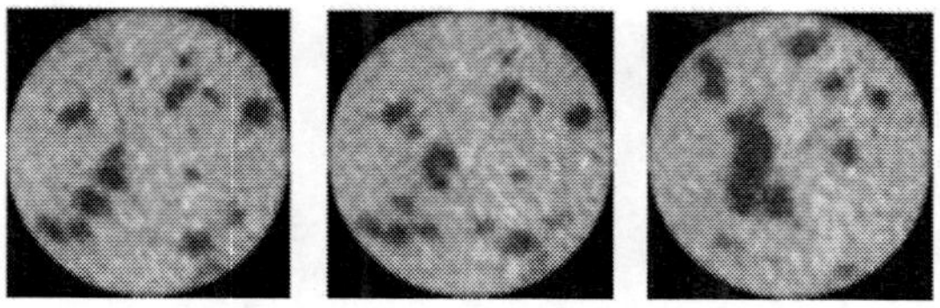

Figure 1. Samples of the reconstructed images of the bubble column's cross-section.

By stacking the images in the time domain, a three-dimensional matrix can be constructed with two spatial dimensions (x and y) and one temporal dimension (z). As shown in Figure 2, by cross examining the matrices of both scanning planes, corresponding bubble objects can be identified. Taking into account the time (Δt) required for a bubble to pass the distance between both planes (Δz), the velocity can be determined with $v = \Delta z / \Delta t$. With the calculated velocity, the actual bubble volumes can be determined by converting the time dimension into the spatial dimension (z).

Results and Discussion

Bubble sizes as well as velocity distributions are constructed across the height of the bubble column by varying the superficial gas velocities. As an example of the results, the bubble size distribution for a superficial gas velocity of 0.02 m/s at the column height of 0.3 m is illustrated in Figure 3. With these results the bubbly flow field within the bubble column can be characterized without any disturbance of measurement probes. Furthermore, this measurement method can be extended towards applications of slurry flows and gas-solid flows.

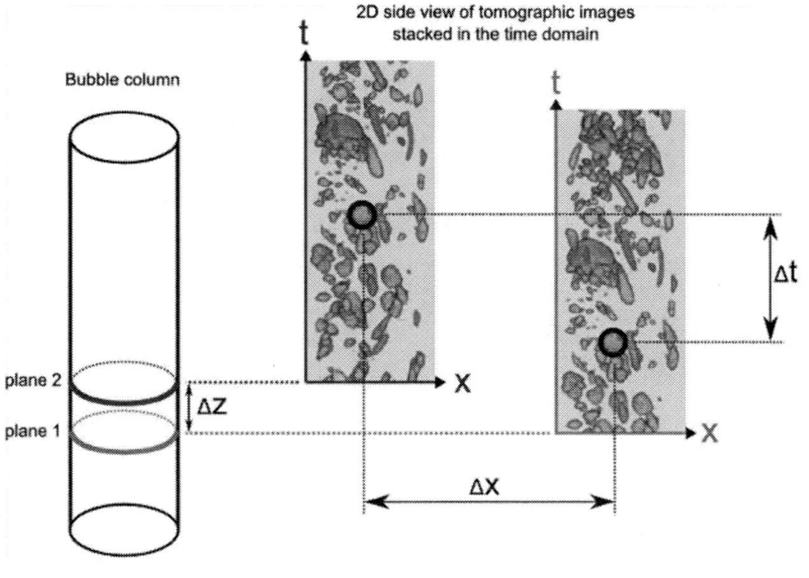

Figure 2. Illustration of X-ray tomographic images reconstructed from the data of the two scanning planes.

References

[1] H.-M. Prasser, Nucl. Eng. Des. 238-3 (2008) 744–770.
[2] F. Fischer, U. Hampel, U, Nucl. Eng. Des. 240–9 (2010) 2254-2259.
[3] S. Rabha, M. Schubert and U. Hampel, Chem. Eng. Sci. 93 (2013) 401-411.
[4] A.C. Kak, M. Slanley, Principles of computerized tomographic imaging, IEEE Press, 1987.

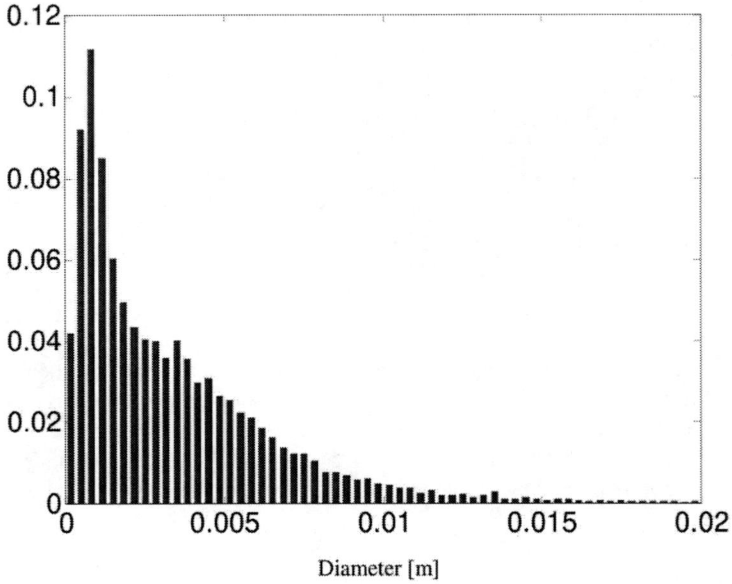

Figure 3. Bubble size distribution for a superficial gas velocity of 0.02 m/s.

New bioreactor designs for the utilization of methane to produce biologically-derived chemicals and fuels using methanotrophs

E Solis (1), J Silverman (1), R Shaver (2), D Arora (2), L Nguyen (1)

(1) Calysta Energy, Inc., Menlo Park, CA
(2) Celanese Corp., Irving, TX

Calysta Energy, Inc. is developing Biological Gas-to-Liquids (BioGTL®) and Biological Gas-to-Chemicals (BioGTC®) platforms that will allow methane-utilizing bacteria (methanotrophs) to effectively produce biologically-derived chemicals and fuels. Methanotrophs contain methane mono-oxygenase (MMO), an enzyme that oxidizes methane into methanol, allowing the bacteria to use an abundant and inexpensive gaseous substrate as a carbon and energy source. There are two major challenges to realizing this technological development: (i) Strain engineering to optimize productivity and overall product yield; (ii) Bioreactor design to maximize methane and oxygen dissolution, maintain safe operation with potentially flammable/explosive conditions, and enable economic process scale-up.

Traditional stir-tank reactors (STR's) provide the industry standard for sugar-based fermentation. However, mass transfer within these units is heterogeneous [1,2] and not easily scalable. Multiple advanced bioreactor designs were developed to reduce power consumption and enhance gas-liquid interactions: Air-lift reactors (ALR's) operate pneumatically to circulate the reaction liquid through a downcomer region and into a riser region, where gas is bubbled through to allow for substrate dissolution and reaction [3,4]. ALR's require less power input than STR's, which depend on mechanical agitation to enhance gas-liquid interaction, but tall units are required to improve single-pass mass transfer. The U-loop reactor [5,6] was developed, scaled-up and operated by BioProtein AS to produce biomass through methanotroph fermentation. Here, the liquid is pumped through a long looped tube, and gaseous substrate is concurrently flowed with the liquid through a series of static mixers to increase mass transfer performance. However, due to the length of the U-loop, capital and pumping costs become inhibitory.

The objective of this project is to develop key bioreactor technology to enable efficient methane fermentation processes for the production of bio-based fuels and chemicals. Using Calysta's in-house fermentation data, we derived the kinetic and mass-transfer parameters for a semi-batch fermentation process of Methylococcus capsulatus (Bath). The kinetic parameters were then used to optimize the conversion of methane to biomass in simulations of new bioreactor designs. The process parameters explored include reactor dimensions, fluid velocities, mass transfer coefficient, liquid recycle fraction, gas hold-up, initial biomass concentration, and pressure. In collaboration with Celanese Corp., we performed computational fluid dynamics (CFD) simulations to predict heat and mass transfer coefficients for different reactor designs. Overall productivity and yields were calculated to compare different bioreactor designs to industry standards and understand how different process parameters affect performance.

References

[1] Sharma C, Malhotra D, Rathore AS. Review of Computational Fluid Dynamics Applications in Biotechnology Processes. *Biotechnology Progress*, 2011, 27, 1497-1510.
[2] Reuss M, Schmalzriedt S, Jenne M. Application of Computational Fluiddynamics (CFD) to Modeling Stirred Tank Bioreactors. In Shugerl K, Bellgardt KH (Ed.), *Bioreaction Engineering*, 2000, 207-246. New York, NY: Springer.
[3] Chisti MY. *Airlift Bioreactors*, 1989, New York, NY: Elsevier.

[4] Chisti Y, Kasper M, Moo-Young M. Mass Transfer in External-Loop Airlift Bioreactors Using Static Mixers. *Canadian Journal of Chemical Engineering*, 1990, 68, 45-50.
[5] Villadsen J. Bioprotein. *Dansk Kemi*, 2002, 83, 20-23.
[6] Olsen DF, Jorgensen JB, Villadsen J, Jorgensen SB. Modeling and Simulation of Single Cell Protein Production. *Proceedings of the 11th Computer Applications in Biotechnology*, 2010, 502-507.

Evaluating Exit Oxygen and Glucose Concentration in an Axial-Flow Bioreactor

Jagdeep T. Podichetty, Kumar Singarapu, and Sundar V. Madihally.
Oklahoma State University

Introduction: The axial-flow bioreactors have recently attracted attention in tissue engineering applications for regenerating a variety of high aspect ratio tissues, such as skin, liver, and bladder. They offer several advantages over conventional perfusion bioreactors such as flow-through and parallel-flow. The nutrient distribution in axial-flow bioreactors is mainly convection-driven, and the ability to operate at high flow rate with relatively low pressure drop and scaffold deformation. The objectives of this study were to validate the outlet oxygen and glucose concentration predictions from computational fluid dynamics (CFD), dispersion model and segregation model using an experimental setup.

Materials and Methods: The axial-flow bioreactor design was evaluated using Computational Fluid Dynamics (CFD) to support 100mm diameter and 2mm thick scaffolds. Chitosan-gelatin scaffolds formed by freeze drying and characterized using scanning electron microscopy (SEM) for pore size and porosity. The elastic properties were determined using Instron 5542 (INSTRON, Canton, MA) for Young's Modulus and a custom-build apparatus for Poisson's ratio values at physiological conditions. The outlet concentration and nutrient distribution were predicted with CFD using the convection diffusion equation with the assumption of uniform cell distribution throughout the scaffold. To predict exit concentrations, the kinetic parameters for Hep G2 were calculated for three different cells densities, which were either half or double of 1.2×10^{12} cells/m3, providing analysis for two doublings. The exit concentrations of oxygen and glucose were also evaluated using segregation model and dispersion model. Bioreactors of appropriate flow configuration were constructed in-house, similar to previous publications. Hep G2 cells were cultured in Eagles's Minimum Essential Medium (EMEM) on chitosan-gelatin scaffolds and were tested for oxygen and glucose consumption. The oxygen concentrations and pressure drop across the scaffold were monitored in real-time. Samples were collected at the outlet to determine glucose concentration. Figure 1 shows the cell culturing process used during the study. Cell seeded scaffolds from 5 different locations in the scaffold were analyzed by histology.

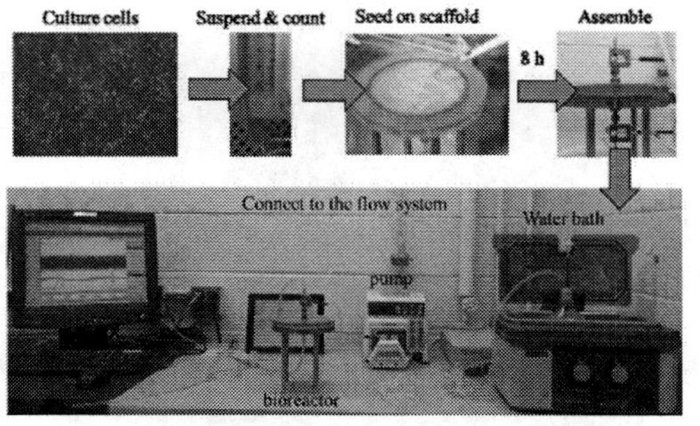

Figure 1: Schematic of cell culture experimental setup with axial-flow bioreactor.

Results and Discussion: The outlet oxygen concentration at 1mL/min with a cell density of 1.2×10^{12} cells/m^3 was 0.166±0.015 mol/m^3 while CFD predicted an outlet concentration of 0.155 mol/m3 (Figure 2). Comparison of the outlet oxygen concentrations between the simulations and segregation model showed good agreement between 15 and 5mL/min flow rates. However, deviations were observed at reduced flow rates of 2mL/min, which increased significantly at 1mL/min. One reason for these deviations could be the extrapolation of the RTD to lower flow rates where diffusion could be dominant rather than convection and this could alter the RTD profile. The dispersion model performed better at lower flowrates. The histology study showed uniform cell distribution in sample obtained from 5 different locations in the scaffold.

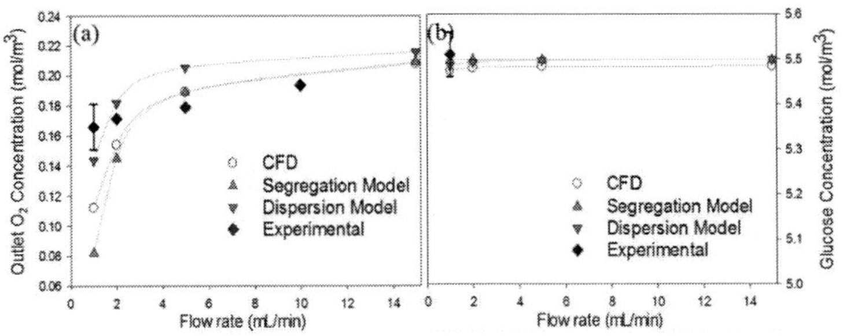

Figure 2: Oxygen and glucose outlet concentration comparison.

Conclusions: Exit concentrations were calculated using CFD, Dispersion, and Segregation model. These models provide an effective way of predicting nutrient consumption at various conditions. These insights help monitor *in vitro* tissue regeneration and help understand the effect of mechanical stimulus on 3D cell culture.

The effect of dispersant on the size distribution of oil droplets from an accidental release: Integration of CFD and PBM

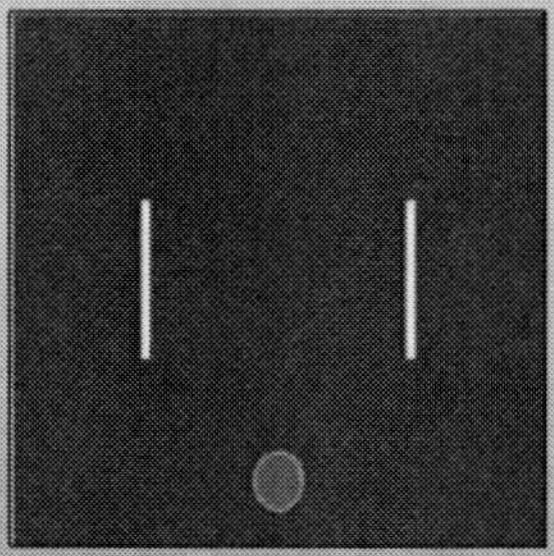

Abhijit Rao
Rupesh K Reddy
Mayur Sathe
K Nandakumar

Tuesday, November 18,2014

Session : Transport of Environmental Contaminants Related to Energy Processes

Outline

- Introduction and Overview
- Numerical Modelling
  - Integration of Multiphase CFD with PBM
- Model Validation
- Results
  - Effect of flow rate on size distribution
  - Effect of dispersant concentration on size distribution
  - Effect of presence of gas phase on size distribution
- Conclusions

Outline

- **Introduction and Overview**
- Numerical Modelling
  - Integration of Multiphase CFD with PBM
- Model Validation
- Results
  - Effect of flow rate on size distribution
  - Effect of dispersant concentration on size distribution
  - Effect of presence of gas phase on size distribution
- Conclusions

Deepwater Oil Spills

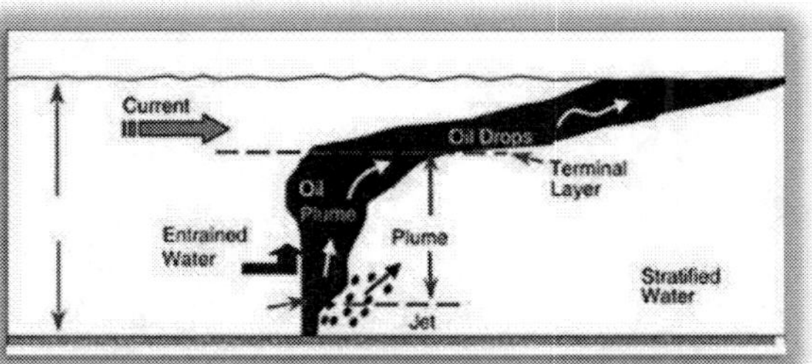

Plume associated with a deepwater oil/gas release (reproduced from Oil in the sea III: Inputs, Fates, and Effects. 2003).

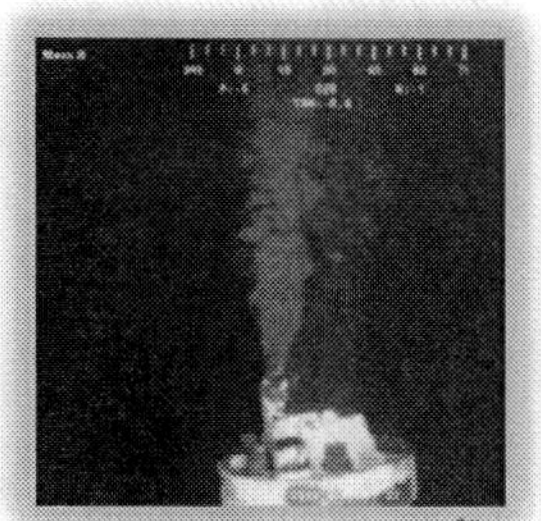

Snapshot off blown out well*#

- Droplets spend significant time in water column.
- Oil / gas mixture – Multiphase Plume
- Chemical dispersants – Oil Spill remediation method

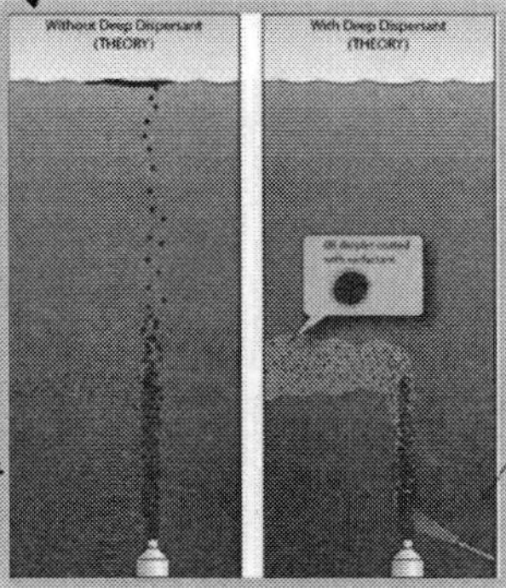

Sub sea Dispersant Application, CE&N, Feb2011

Droplet size distribution in a subsurface release helps in
- Determining fate of oil droplets.
- Estimation of effective interfacial area for dissolution process.

Action of dispersants

- Dispersant : Surfactant +Solvent
- Surfactants lower the interfacial tension at oil-water interface.
- Local turbulence assists droplets disintegration

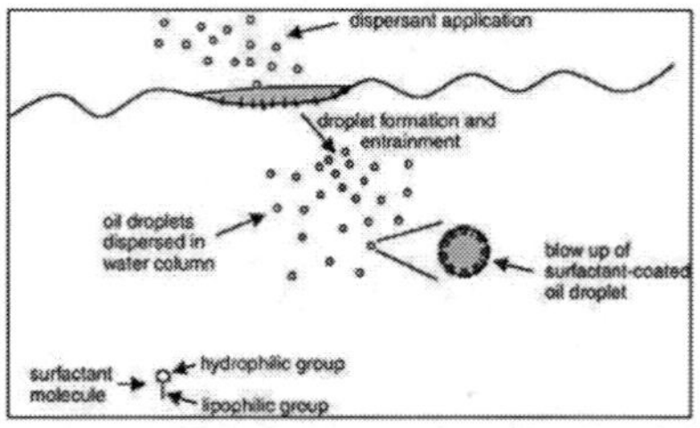

Surface dispersant application on the oil slicks#

Turbulent plume characteristics and Droplet Generation

- **ZFE (Zone of flow establishment)**
  - Flow is momentum dominated.
- **ZEF (Zone of established flow)**
  - Flow is buoyancy driven.
- **Droplets form by turbulent breakup in ZFE**
- **Droplet size distribution in rising plume stabilizes quickly** as turbulence diminishes significantly.

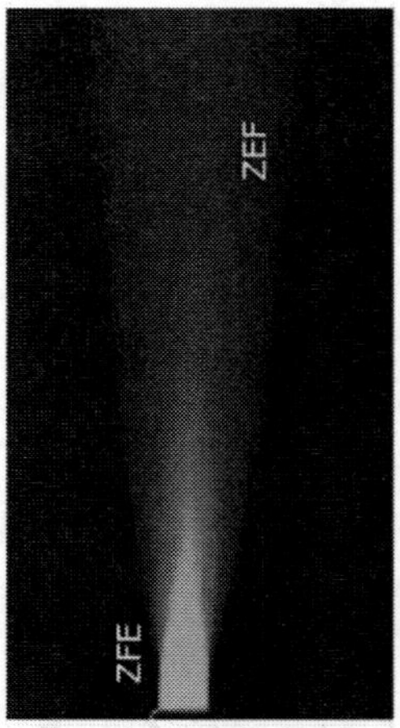

Zones in a turbulent plume

 # "Fate of Dispersants Associated with the Deepwater Horizon Oil Spill", Kujawinski et al, Environ. Sci. Technol (2010).

Droplet dynamics in surfactant laden turbulent environment

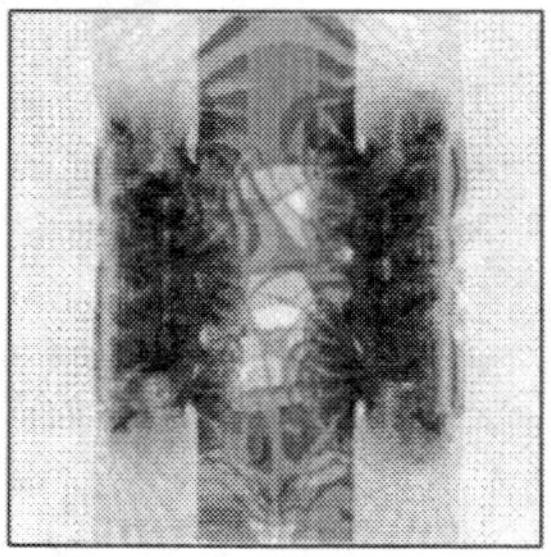

Velocity vectors

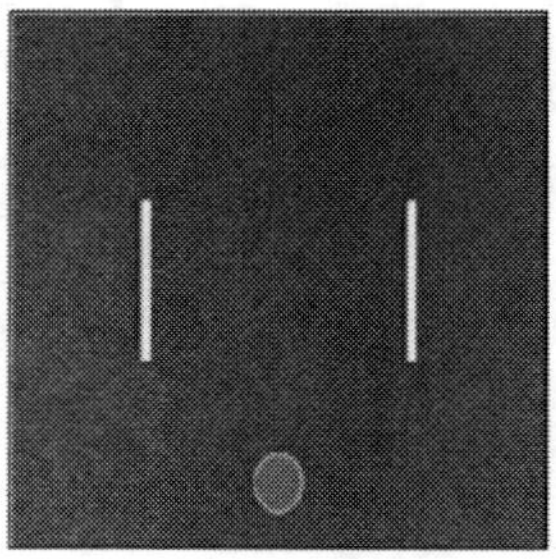

Contour plot of oil phase vof

Droplet in a turbulent field

Jet released in surfactant laden environment

Droplet disintegration in a turbulent field

- Collision with small eddies
- Energy of approaching eddy overcomes the surface energy of droplet
- Balance between dynamic pressure τ_i surface stress τ_s dictates breakup

$d_{p,i}$

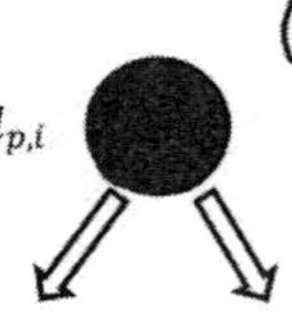

$d_{p,j}$ $\quad (d_{p,i}^{\;3} - d_{p,j}^{\;3})^{1/3}$

$$We_{cr} = \frac{\tau_i}{\tau_s} = \frac{\rho_c v_c'^2 \pi d^2}{\sigma \pi d} \qquad v_c' = 2(l_{ec}\varepsilon_c)^{2/3}$$

Outline

- Introduction and Overview
- **Numerical Modelling**
  - **Integration of Multiphase CFD with PBM**
- Model Validation
- Results
  - Effect of flow rate on size distribution
  - Effect of dispersant concentration on size distribution
  - Effect of presence of gas phase on size distribution
- Conclusions

Integrated Model

- Multiphase model (Mixture Model)
  - Accounts for the dispersed and continuous phases
- Turbulence model (k-ε Model)
  - Predicts the effect of turbulence on flow field.
- Population Balance Model (Discrete Method)
  - Accounts for droplet breakage events and yields size distribution.

Assumptions

- Newtonian fluids.
- Turbulence is isotropic.
- Breakup dominant process (Plume spreading).

Mixture Model (Multiphase model)

- Eulerian approach.
- Phases treated as interpenetrating continuum.
- Computationally less intensive.

Continuity Equation :

$$\frac{\partial(\rho_m)}{\partial t} + \nabla \cdot (\rho_m \overrightarrow{v_m}) = 0$$

$\overrightarrow{v_m}$ is mass averaged velocity

$$\overrightarrow{v_m} = \frac{\sum_{k=1}^{n} \alpha_k \rho_k \overrightarrow{v_k}}{\rho_m} \qquad \rho_m = \sum_{k=1}^{n} \alpha_k \rho_k$$

Momentum Equation:

$$\frac{\partial}{\partial t}(\rho_m \overrightarrow{v_m}) + \nabla \cdot (\rho_m \overrightarrow{v_m} \overrightarrow{v_m}) = -\nabla p + \nabla \cdot \left[\mu_m\left(\nabla \overrightarrow{v_m} + \nabla \overrightarrow{v_m}^T\right)\right] + \rho_m \vec{g} + \vec{F} + \nabla \cdot \left(\sum_{k=1}^{n} \alpha_k \rho_k \overrightarrow{v_{dr,k}}\, \overrightarrow{v_{dr,k}}\right)$$

where $\overrightarrow{v_{dr,k}} = \overrightarrow{v_k} - \overrightarrow{v_m}$ drift velocity

Evaluates v_m, p and α_k

Volume fraction Equation:

$$\frac{\partial(\alpha_k \rho_k)}{\partial t} + \nabla \cdot (\alpha_k \rho_k \overrightarrow{v_m}) = -\nabla \cdot (\alpha_k \rho_k \overrightarrow{v_{dr,k}})$$

k-ε mixture turbulence model

- RANS (Reynolds Averaged Navier Stokes) approach.
- Turbulent viscosity is evaluated
- Focuses on mechanisms which affect turbulent kinetic energy.

$$\boldsymbol{k = \frac{1}{2}(u'^2 + v'^2 + w'^2) \qquad \varepsilon = 2\vartheta \overline{s_{ij}' \cdot s_{ij}'}}$$

where $s_{ij}' = \left(\nabla \overrightarrow{v_m} + \nabla \overrightarrow{v_m}^T\right)$

k Equation (turbulent kinetic energy)

$$\frac{\partial(\rho_m k)}{\partial t} + \nabla \cdot (\rho_m \overrightarrow{v_m} k) = \nabla \cdot \left(\left(\frac{\mu_{t,m}}{\sigma_k}\right) \nabla k\right) + G_{k,m} - \rho_m \varepsilon$$

ε Equation (turbulent dissipation rate)

$$\frac{\partial(\rho_m \varepsilon)}{\partial t} + \nabla \cdot (\rho_m \overrightarrow{v_m} \varepsilon) = \nabla \cdot \left(\left(\frac{\mu_{t,m}}{\sigma_\varepsilon}\right) \nabla \varepsilon\right) + \frac{\varepsilon}{k}(C_{1\varepsilon} G_{k,m} - C_{2\varepsilon} \rho_m \varepsilon)$$

where $\mu_{t,m} = \rho_m C_\mu \frac{k^2}{\varepsilon}$;

$G_{k,m} = \mu_{t,m}\left(\nabla \overrightarrow{v_m} + \nabla \overrightarrow{v_m}^T\right) : \nabla \overrightarrow{v_m}$ (Production of turbulent kinetic energy)

Solves for μ_t using k and ε

C_μ=0.09 ; σ_k=1 ; σ_ε=1.2; $C_{1\varepsilon}$=1.44 ; $C_{2\varepsilon}$=1.92

Population Balance Model

Number density function is given by :

$$\frac{\partial}{\partial t}[n(V,t)] + \nabla\cdot[\vec{u}n(V,t)] + \nabla_v\cdot[G_v n(V,t)]$$

Convection Growth

$$= \frac{1}{2}\int_0^V \Omega_{br}(V-V',V')\,n(V-V',t)n(V',t)dV' - \int_0^\infty \Omega_{br}(V,V')n(V,t)n(V',t)dV'$$

Birth due to coalescence Death due to coalescence between

$$+\int p\Omega_{br}(V')\beta(V|V')n(V',t)dV' - \Omega_{br}(V)n(V,t)$$

Birth due to breakage Death due to breakage

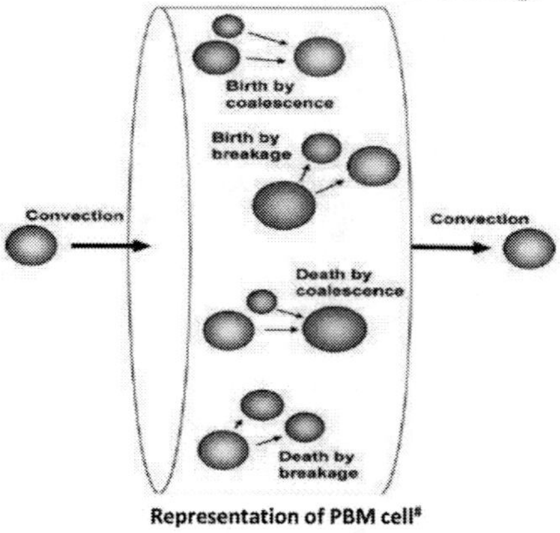

Representation of PBM cell#

#F.Azizi, A.M.A Taweel, Chemical engg Journal ,166(2011), 715-725

Solves for $n(V,t)$ using σ, ϑ_T and ε

Methods

- **Discrete Method**
- Standard Method of Moments
- Quadrature Method of Moments
- Direct quadrature Method of Moments

Discrete Method

- Discretization of PSD into series of number of bins of discrete sizes.

$$V_i = V_{min}q^{i-1}$$

V_{min} and q are specified

- For each bin, following eqn is solved

$$\frac{\partial}{\partial t}(\rho_s\alpha_i) + \nabla\cdot(\rho_s\alpha_i u_s) + \frac{\partial}{\partial V}\left(\frac{G\rho_s\alpha_i}{V}\right) = \rho_s V_i\left(B_{ag,i} - D_{ag,i} + B_{br,i} - D_{br,i}\right)$$

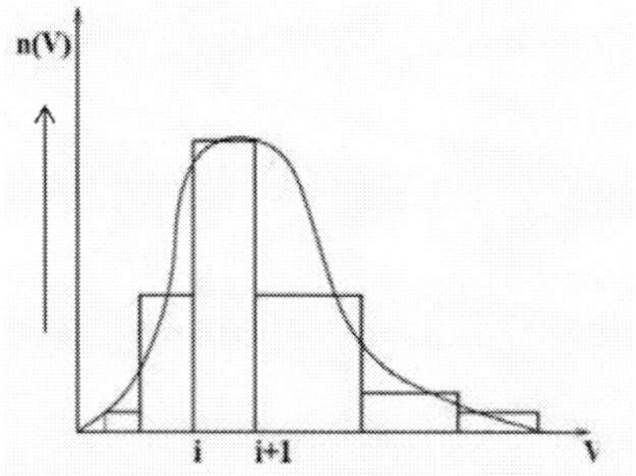

Breakage Kernel - Lehr Model (Lehr and Mewes ,2002)

$$\Omega_{br} = b(v_i) * P(v_j, v_i)$$

Breakage rate Breakup frequency Daughter size distribution

$$\Omega_{br}(V,V') = K\int_{\xi_{min}}^{1}\frac{(1+\xi)^2}{\xi^{11/3}}exp\left(-b\xi^{-11/3}\right)d\xi$$

$$K = \frac{1.19\sigma}{\rho f^{1/3}\varepsilon^{1/3}d^{7/3}} \qquad b = \frac{2We_{cr}\sigma}{\rho\varepsilon^{2/3}d^{5/3}f^{1/3}} \qquad \xi = \frac{\lambda}{d}$$

λ= Eddy size ; d = droplet ; We_{cr} = Critical Weber Number ;

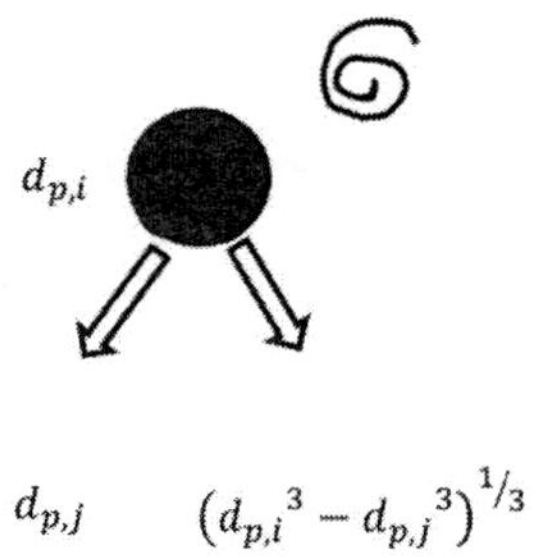

Ω_{br}= f(ε,d, σ, We_{cr})

Integration of CFD and PBM

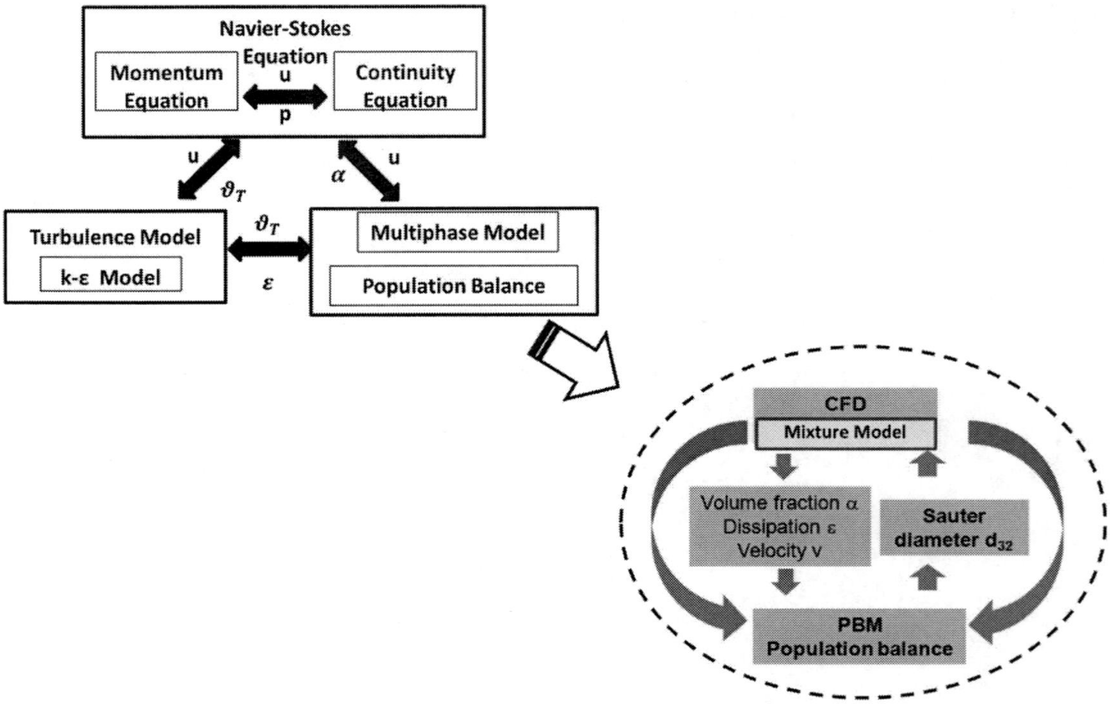

Outline

- Introduction and Overview
- Numerical Modelling
  - Integration of Multiphase CFD with PBM
- **Model Validation**
- Results
  - Effect of flow rate on size distribution
  - Effect of dispersant concentration on size distribution
  - Effect of presence of gas phase on size distribution
- Conclusions

SINTEF Meso Scale Experiment#

- Tank capacity 40m^3 (Height: 6m ; Diameter: 3 m)
- Droplet size distribution measured using LISST (Laser diffractometer)

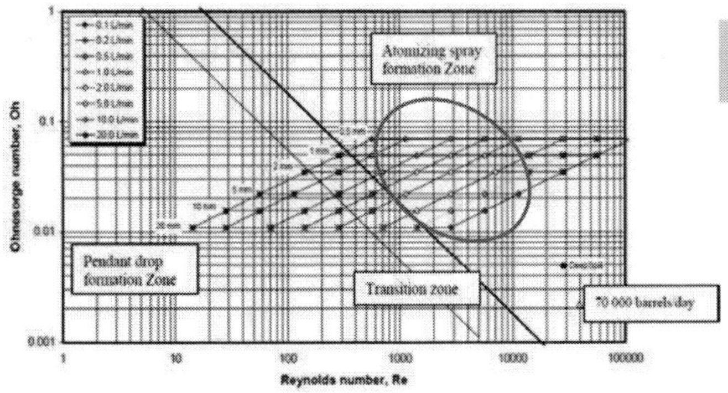

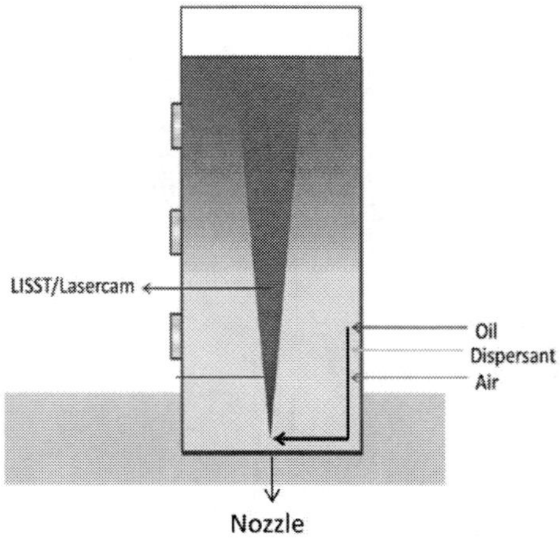

Droplet size distribution for

- Flow Experiments.
- Experiments at different Dispersant to Oil ratio.

 # Johansen Ø, Brandvik PJ, Farooq U. Droplet breakup in subsea oil releases – Part 2: Predictions of droplet size distributions with and without injection of chemical dispersants. Marine Pollution Bulletin. 8/15/ 2013;73(1):327-335.

Simulation details

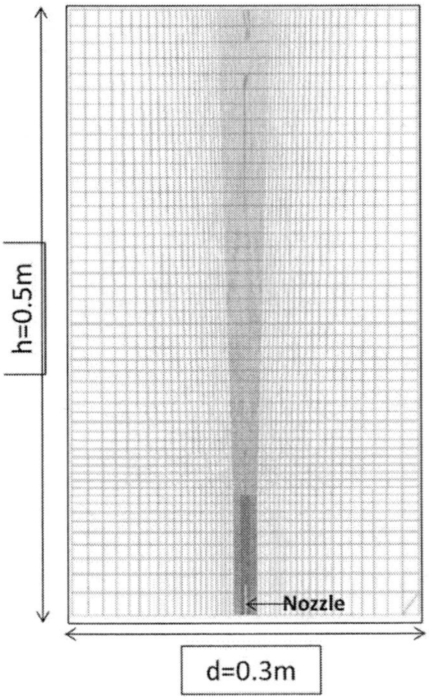

Computational domain (central plane)

	Density(kg/m3)	**Viscosity (kg/m.s)**
Dispersed Phase		
Oil	839.3	0.005
Continuous Phase		
Water	999.5	0.001
Dispersant	Corexit 9527 A	

Turbulence Model	k-ε realizable (RANS)
PBM Method	Discrete Method
Number of bins	30
Minimum bin size (m)	4.5×10^{-6}
Maximum bin size (m)	4.6×10^{-4}
Breakage Kernel	Lehr Model
Computational cells	200,000

Model Validation

Flow Rate (L/min)	1.5
IFT (mN/m)	15.5
Nozzle diameter (mm)	1.5

$$We_{cr} = 0.24$$

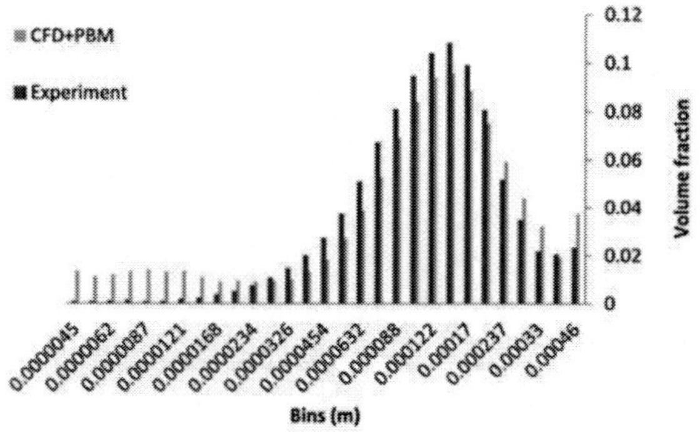

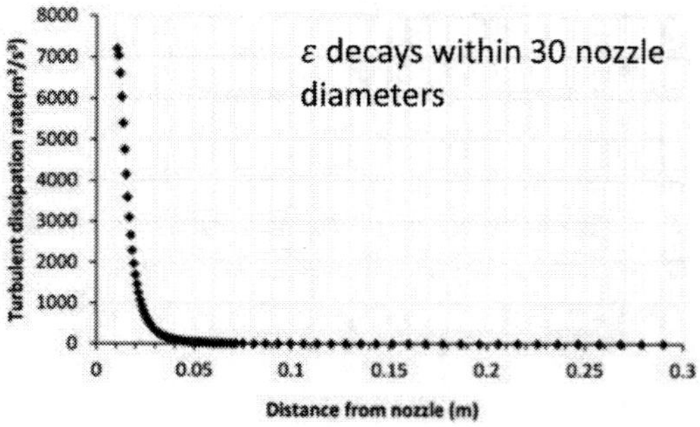

Outline

- Introduction and Overview
- Numerical Modelling
  - Integration of Multiphase CFD with PBM
- Model Validation
- **Results**
  - **Effect of flow rate on size distribution**
  - **Effect of dispersant concentration on size distribution**
  - **Effect of presence of gas phase on size distribution**
- Conclusions

A) Effect of Flow Rate

Flow Rate (L/min)	0.5, 1.2 and 2.8
Interfacial Tension (mN/m)	15.5
Nozzle diameter (mm)	1.5

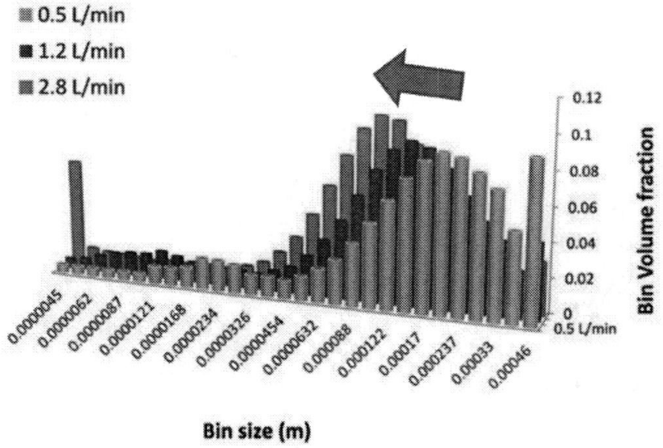

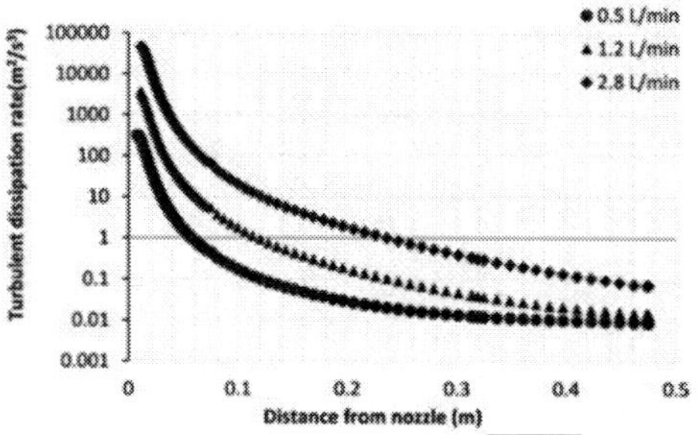

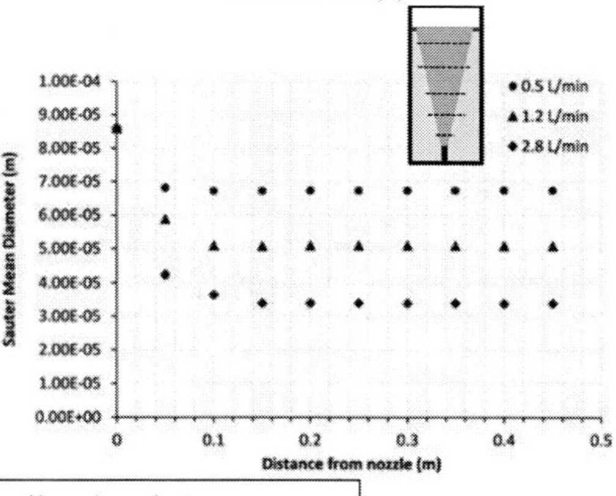

Higher flow rate yields the smaller droplets

B) Effect of Dispersant concentrations on droplet size distribution

DOR	No dispersant	1: 250	1:25
IFT (mN/m)	15.5	2.6	0.05
Flow Rate (L/min)	1.5		
Nozzle diameter (mm)	1.5		

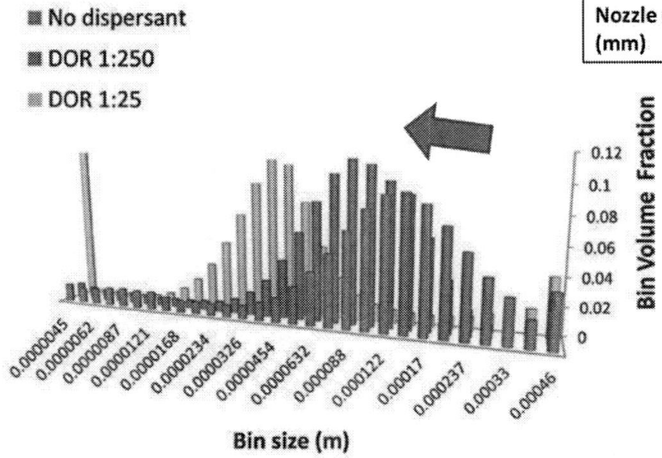

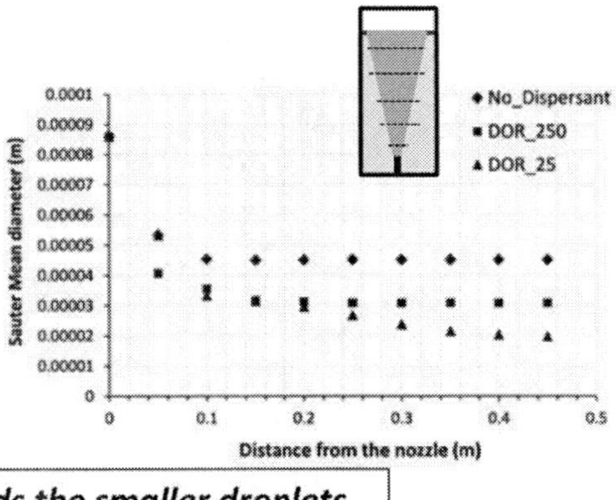

Higher dispersant concentration yields the smaller droplets

C) Effect of presence of Air* on droplet size distribution

Flow Rate (L/min)	1.5
Interfacial Tension (mN/m)	15.5
Nozzle diameter (mm)	1.5

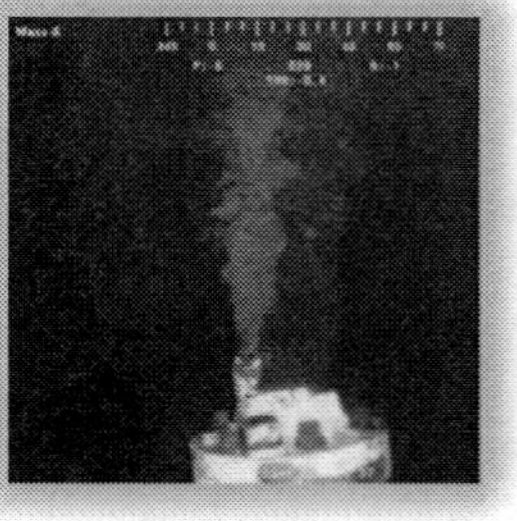

Snapshot off blown out well

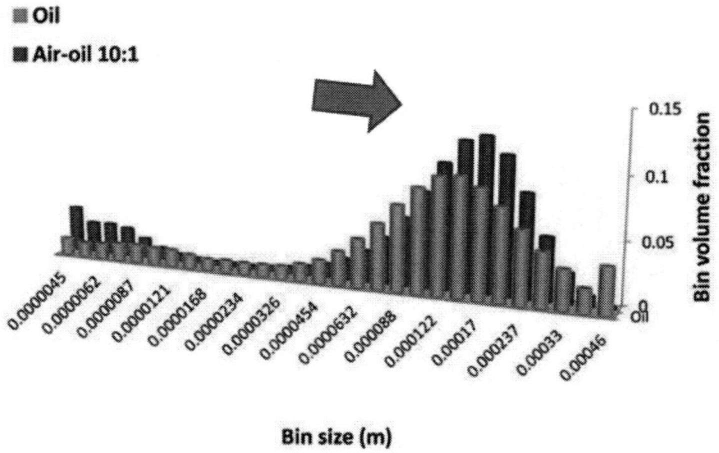

- Collision between bubbles and droplets has not been considered.
- Presence of air makes plume more buoyant.
- Average time spent by droplet in region of high turbulence is small.

* Typical gas to oil ratio in DWH varied from 1500 to 2500 ft³/barrel oil (350 to 600)

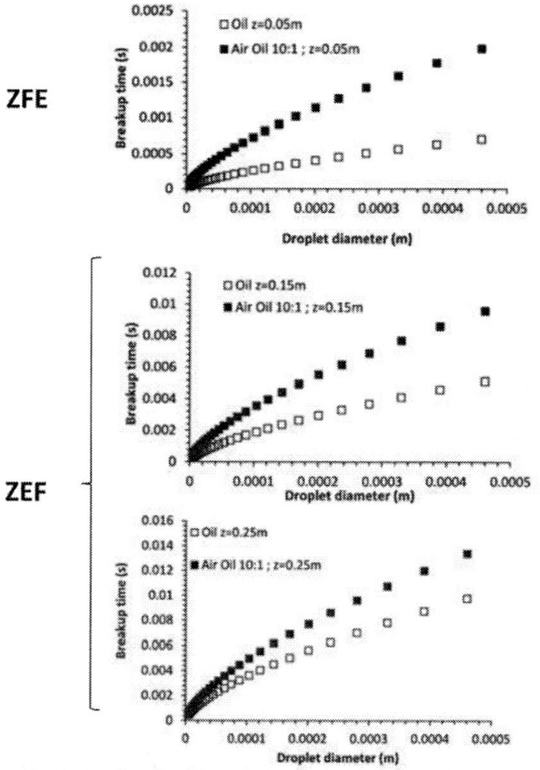

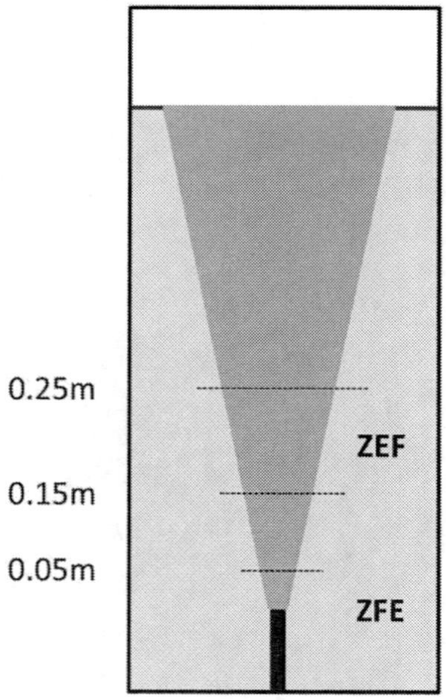

$$t_{br} \approx \frac{d^{2/3}}{\varepsilon^{1/3}}$$

LSU

Conclusions

- Integration of multiphase CFD model with PBM to predict droplet size distribution.
- Increase in flow rate of oil phase yields smaller sized droplets.
- Decrease in DOR shifts size distribution to smaller bins.
- Presence of gas phase reduces the residence time of oil phase in ZFE and yields larger droplets.

LSU

Acknowledgements

- Gulf of Mexico Research Initiative
- High Performance Computing at LSU
- Louisiana Optical Network Initiative
- Extreme Science and Engineering Discovery Environment
- EPIC group at LSU

Questions!!!

Morphology and release kinetics parameters of mercury in coal gangue

Libao Yin[1], Peixin Wu[2], Zhengyang Gao[2]

1 Electric Power Research Institute of Guangdong Power Grid Corporation, Guangzhuo, 510000, Guangdong Province, China

2 School of Energy and Power Engineering, North China Electric Power University, Baoding 071003, Hebei Province, China)

ABSTRACT: Three different regions of gangue were selected and analyzed in this study to examine the characteristics of mercury (Hg) emission in the coal gangue combustion process. The effect of temperature on the properties of Hg emissions from coal gangue, including the release characteristics of Hg under different final temperatures and different heating temperatures under different heating times, were investigated. The kinetic parameter of Hg during the heating process was calculated. Finally, the characteristics of Hg emission in the combustion process was examined to study the impact of particle fineness and residence time as well as coal gangue combustion with coal on the Hg emission characteristics. Experimental results show that the release of Hg is determined by the initial content of Hg in coal gangue and the Hg occurrence form. At the same temperature, the amount of Hg released from coal gangue increases initially and then decreases with time. The Hg heat reaction series in the process of coal gangue is 1. The adsorption amount of Hg at fly ash after gangue combustion is higher than that in pulverized coal. The adsorption capacity of Hg in fly ash after coal gangue burning decreases significantly with the increase in wind.

KEYWORDS: coal gangue; mercury; temperature; combustion

0 Introduction

Coal is the primary energy source in our country. Coal mining and washing can produce a large amount of coal gangue. The output of coal gangue, which is generally 10% greater than that of raw coal production, can be as high as 20%[1]. The current amount of coal gangue piled up in our country is approximately more than 4.5 billion tons; this figure is the largest amount of industrial solid waste in emissions and heap memory[2]. Coal gangue contains large amounts of toxic trace elements. Coal gangue typically consists of 10% to 25% of carbon fuel and a certain amount of pyrite. Spontaneous combustion occurs in sulfur content above 1% that of coal gangue[3]. Pyrite has a low ignition point of approximately 290 °C. When the temperature reaches the ignition point of pyrite, pyrite spontaneously releases a large amount of heat and thus causes carbon combustion[3]. In the spontaneous combustion of coal gangue, the surface temperature is as high as 300 °C to 400 °C; the internal temperature can reach up to 800 °C to 1,000 °C[4]. Such high temperature causes coal gangue to be released to the environment as pollutants.

Mercury (Hg) is one of the toxic trace elements in coal and gangue[5], with mainly mercuric sulfide, organic Hg, and elemental Hg forms combined with sulfides (pyrite and sphalerite) and selenite[6, 7]. At present, most research on fuel Hg release focuses on the discharge and control of Hg after combustion. The presence and formation of Hg in coal gangue [8, 9] and the release of Hg with a temperature increase require in-depth research. The reaction of the release kinetics parameters of Hg in coal gangue has not been reported.

To this end, a release characteristic test for Hg in coal gangue was conducted in this study under different temperatures with three types of coal gangue at different heating temperatures. The principle of Hg release with temperature change in three types of coal gangue was determined based on the residue of Hg in coal gangue after heating. The occurrence of Hg in coal gangue was analyzed to calculate the reaction of the release kinetics parameters of Hg in the three types of coal gangue[10, 11].

1 Test method and equipment

1.1 Test equipment and condition

The experimental coal gangue samples were Handan Wuan mine gangue, Shanxi Hunyuan coal gangue, and Wuhai coal gangue labeled as NO. 1, NO. 2, and NO. 3, respectively. The element and industrial analyses of the experimental coal gangue are shown in Table 1. The data in the table show that the volatile content in the three types of coal gangue is basically similar; the C content in sample NO. 3 is significantly higher than that in samples NO. 1 and NO. 2. The Hg content in the three types of coal gangue is 240, 692, and 508 ng/g, respectively. The Hg content in NO. 1 coal gangue is the average level of the national Hg content. In NO. 2 and NO. 3, the Hg content is higher than the average level of the national Hg content.

Table 1. Analysis of the coal samples

	Industrial analysis (received)			Elemental analysis (received)				
	M_{ad}	A_{ad}	V_{ad}	C_{ad}	H_{ad}	O_{ad}	N_{ad}	S_{ad}
1	0.90	81.78	10.45	8.44	1.30	6.10	0.22	1.36
2	1.10	78.58	10.91	9.74	1.41	5.80	0.35	2.92
3	0.78	65.14	10.66	24.5	1.86	5.70	0.62	1.50

First, the three samples were air-dried naturally under laboratory conditions, ground, and screened with a 200 mesh sieve. Each 1 g of weighing precision is equal to 0.1 mg in all the three samples. An electric heating furnace was utilized to heat the samples to

100 °C, 200 °C, 300 °C, 400 °C, 500 °C, and 800 °C. The quality and Hg content of the samples were measured after heating. The test was repeated three times, and the average was obtained.

Milestone Company's DMA80 direct Hg measurement instrument was utilized for Hg content analysis before and after heating. The analyzer was first calibrated with standard samples to guarantee the accuracy of the analysis. The Hg content of the standard samples was 500 ng/g, the result of the analysis is 520 ng/g, and the correction coefficient is 25/26.

2 Examination results and analysis

2.1 Coal gangue weightlessness characteristics at different terminal heating temperatures

Figure 1 shows the weightlessness rate of the three types of coal gangue at different terminal heating temperature curves. The ratio of the lost weight and the initial weight of coal gangue before and after heating is the weightlessness rate. The figure shows that the curve of the weightlessness rate of the three types of coal gangue is basically similar after 2 h under the conditions of heat finishing temperature of 800 °C and heating temperature of 500 °C. At a terminal heating temperature of over 500 °C, the three types of coal gangue can burn completely in 2 h.

When heated at final temperatures below 500 °C, the features of the weightlessness rate of the three types of coal gangue are not completely similar. NO. 1 and NO. 2 are weightless when heated. When the heating terminal temperature is 100 °C and 200 °C, the weightlessness rate is less than 1%. When the heating terminal temperature is in the 300 °C to 500 °C range, the weightlessness rate significantly increases with the increase in temperature.

Data show that the regularity of NO. 3 coal gangue is different from that of NO. 1 and NO. 2. When the terminal heating temperature is 100 °C, NO. 3 coal gangue, compared with before heating, exhibits a trace of weightlessness. However, when heated at high temperatures of 200 °C and 300 °C, NO. 3 coal gangue becomes heavy instead. The heating weight increases by 0.013% when the terminal temperature is 200 °C and increases by 0.08% when the terminal temperature is 300 °C. The weight increased because NO. 3 coal gangue adsorbed oxygen (O) in the air. The data in Table 1 show that the C content of NO. 3 coal gangue is high. The activities of C and O in the air in NO. 3 coal gangue included the slow reaction at low heating temperature and adsorption on coal gangue activity. O and C reacted to increase the sample weight.

Table 2. Thermal dissociation temperatures for Hg compounds[12]

Hg compound	Pyrolysis peak (°C)	Peak value (°C)
HgCl2	70–220	120 ± 10
Hg2Cl2	60–220	80 ± 5; 130 ± 10
HgBr2	60–220	110 ± 5
Black HgS (black sand)	170–290	205 ± 5; 245 ± 5
Red HgS (cinnabar)	240–350	310 ± 10
HgSO4	500–600	540 ± 20
HgO	430–560	505 ± 5

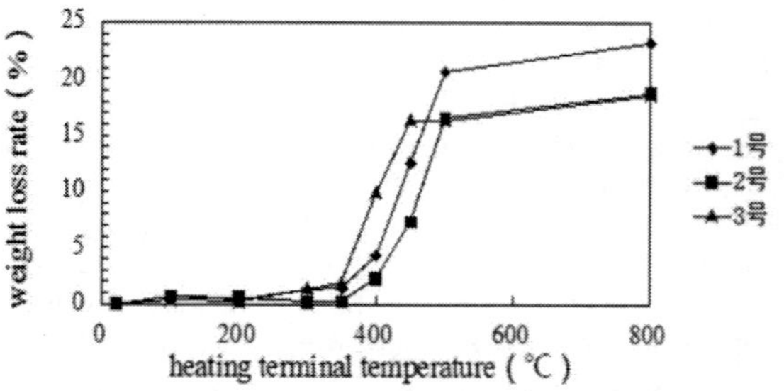

Figure 1. Gangue weight loss rate at different terminal heating temperatures

2.2 Coal gangue Hg release characteristics at different terminal heating temperatures

Figures 2 and 3 show the Hg concentrations in the samples at different terminal heating temperatures as well as the Hg residue rate curves. Figure 2 shows that when the temperature is below 100 °C, the Hg content in the samples basically remain unchanged. Compared with the temperature range in Figure 1, no significant mass loss was observed in coal gangue. Coal gangue is stable at low temperatures of Hg. When the temperature exceeds 100 °C, the Hg content of the coal gangue samples began to decrease. At 200 °C, the Hg concentrations of NO. 1 and NO. 3 coal gangue samples decreased by approximately 10% compared with unheated coal gangue. NO. 2 coal gangue decreased by approximately 30%. The weightlessness rates of the three coal gangue samples were not significant at 200 °C. At this time, the release of Hg is not related to the release of quality. The NO.1 and NO. 3 coal gangue curve slopes in Figure 2 increased when the temperature is increased. This result indicates that the Hg release rate increases when the temperature increases; when the temperature exceeds 400 °C, the curve slope is reduced and the Hg release rate is decreased.

Figure 3 shows the residual rate curves of the three types of coal gangue at a similar heating temperature. Overall, the Hg residue rate of NO. 3 coal gangue is the highest and that of NO. 2 coal gangue is the lowest. This result suggests that the heat stability of Hg in the NO.2 coal gangue is the worst. The heat stability of Hg in NO. 3 coal gangue is the most stable. The geometrical shapes of the three types of coal gangue in Hg are different. As shown in Figures 2 and 3, in the range of 100 °C to 500 °C, the Hg release quantity of the three types of coal gangue is different. The Hg release quantity of NO. 2 coal gangue has two peaks at the 100 °C to 150 °C and 300 °C to 400 °C temperature sections. In the 100 °C to 150 °C temperature section, the chemical forms of Hg are HgCl2, Hg2Cl2, HgBr2, and HgS. In the 300 °C to 400 °C temperature section, the chemical forms of Hg are HgS and Hg2SO4. The Hg release quantities in NO. 1 and NO. 3 coal gangue have one peak in the 350 °C to 400 °C temperature section, and the chemical form of Hg is Hg2SO4.

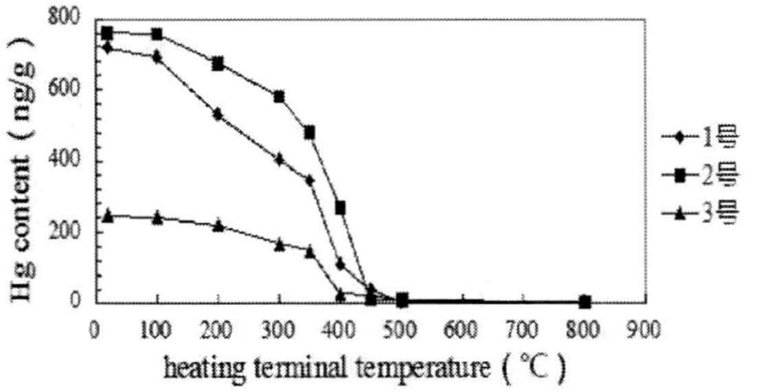

Figure 2. Different heating temperatures for the rest of the Hg content

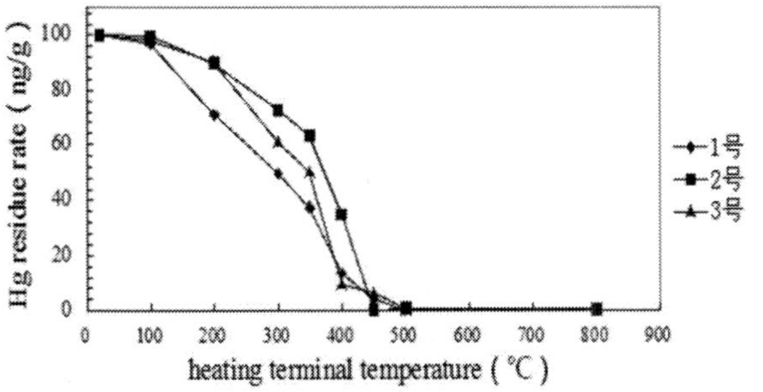

Figure 3. Hg residue rate under different terminal heating temperatures

Figure 4 shows the Hg emission curve of the three types of coal gangue at different temperatures. The data in the graph show that at the same heating temperature, NO. 1 coal gangue releases the highest amount of Hg because the Hg content in NO. 1 coal gangue is the highest and the Hg heat stability is the worst; thus, a large amount of Hg is released and environmental pollution is at its worst. For the NO. 2 and NO. 3 coal gangue curves, although Hg content of NO. 2 coal gangue is far less than that of NO. 3 coal gangue, the Hg release quantity in the two coal gangue is basically similar when the heating temperature is below 300 °C. The causes of this phenomenon can be seen from the curve shown in Figure 3. The curve in Figure 3 shows that at a heating temperature below 400 °C, the Hg residue rate of NO. 3 coal gangue is greater than that of NO. 2. The Hg emissions in the heating process of different gangue depend on two aspects; one is the original Hg content in the coal gangue, and the other is the Hg release characteristics of gangue, which is released by the reaction kinetics parameters.

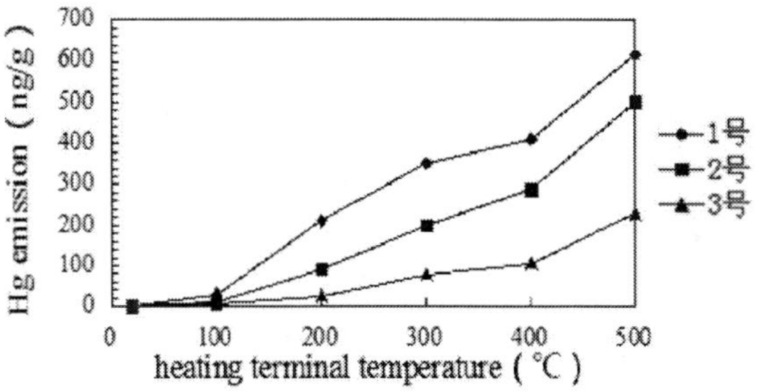

Figure 4. Hg emission at different terminal heating temperatures

2.3 Hg release reaction kinetics parameters of the three types of coal gangue

The previous analysis indicates that the key factor that affects Hg release is the Hg release reaction kinetics parameters of the heating process. Given the low Hg content in coal gangue, the reaction kinetics parameters of the three types of coal gangue cannot be obtained by a thermogravimetric experiment. A muffle furnace was used in the test process. For every 5 °C increase in temperature starting from 100 °C, the corresponding time was recorded. For every 50 °C increase, the sample was removed and time and the final measurement of the sample quality and mercury levels were obtained after heating.

As observed from the experimental process, the reaction process description should adopt the variable-temperature no-heterogeneous reaction kinetics equation[13,14], which is

$$\frac{d\alpha}{dt} = k(T)f(\alpha) = A \cdot \exp[-E/(RT)] \cdot f(\alpha), \quad (2\text{-}1)$$

where $\beta = dT/dt$ is the heating rate; hence,

$$\frac{d\alpha}{dt} = \frac{A}{\beta} \cdot \exp[-E/(RT)] \cdot f(\alpha), \quad (2\text{-}2)$$

$$f(\alpha) = (1-\alpha)^n. \quad (2\text{-}3)$$

Substituting (2-3) into (2-2), we obtain

$$\frac{d\alpha}{(1-\alpha)^n} = \frac{A}{\beta} \cdot \exp[-E/(RT)]dT. \quad (2\text{-}4)$$

The integral method was adopted with order $u = \frac{E}{RT}$, plug in, and integral

$$\int_0^{\alpha} \frac{d\alpha}{(1-\alpha)^n} = \frac{AE}{\beta R} \int_{\infty}^{u} \frac{-e^{-u}}{u^2} du = \frac{AE}{\beta R} p(u), \quad (2\text{-}5)$$

$$\begin{aligned}p(u)&=\int_0^u -\frac{\exp(-u)}{u^2}du\\&=\int_0^u \frac{1}{u^2}d\exp(-u)\\&=\frac{\exp(-u)}{u^2}-\int_\infty^u \exp(-u)(-2)u^{-3}du\\&=\frac{\exp(-u)}{u^2}(1-\frac{2!}{u}+\frac{3!}{u}-\frac{4!}{u}+\ldots).\end{aligned}$$

Omit the second and later term:

$$p(u)=e^{-u}\left(\frac{1}{u^2}\right)\left(1-\frac{2}{u}\right) \quad (2\text{-}6)$$

According to the Coats–Redfern method, when we directly integrate variables into Equation (2-5) and obtain the logarithm of both sides, we obtain

$$\ln\left[\frac{1}{T^2}\int_0^\alpha \frac{d\alpha}{(1-\alpha)^n}\right]=\ln\left[\frac{AR}{\beta E}\right]-\frac{E}{RT}. \quad (2\text{-}7)$$

For different reaction models, n acquires different values.

When n = 1, the reaction is a first-order reaction.

$$\ln\left[\frac{-\ln(1-\alpha)}{T^2}\right]=\ln\frac{AR}{\beta E}-\frac{E}{RT}. \quad (2\text{-}8)$$

When $n \neq 1$, we derive the integral result as

$$\ln\left[\frac{1-(1-\alpha)^{1-n}}{T^2(1-n)}\right]=\ln\frac{AR}{\beta E}-\frac{E}{RT}. \quad (2\text{-}9)$$

With Equations (2-8) and (2-9) on the left for Y, $X=-1/T$, a straight line is obtained with the linear regression equation. Kinetic parameter E, which refers to the former factor A, is obtained by the slope and intercept. The value of n is uncertain. Thus, we must assume the n values, with each value of n fitting a straight line. The linear correlation peak value of n is applied as the true value of n. After determining the value of n, the slope of this straight line and the intercept can be calculated as E and A. When n is different, the fitting equation of the straight line and the correlation coefficient also differ as shown in Figure 5.

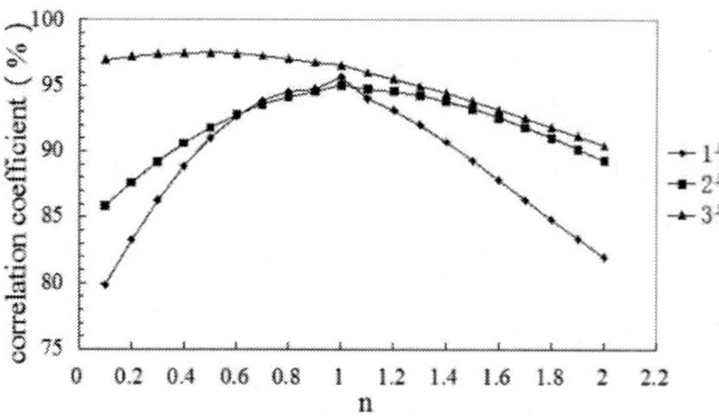

Figure 5. Correlation coefficient R^2 with the change in the value of n

As shown in Figure 5, the correlation coefficient initially increases with the increase in the value of n and then decreases. The maximum of NO. 1 and NO. 2 samples is obtained when n = 1, and the maximum of NO. 3 samples is obtained when n = 0.5. Figures 6 and 7 show the reaction kinetics parameter curves of the three samples when n = 0.5 and when n = 1. As shown in Figure 6, when n = 0.5, the maximum correlation coefficient of the NO. 3 samples is 0.9719, whereas the correlation coefficients of the NO. 1 and NO. 2 samples are 0.9048 and 0.9245, respectively. As shown in Figure 7, when n = 1, the maximum correlation coefficients of NO. 1 and NO. 2 are 0.9410 and 0.9626, respectively, and that of the NO. 3 sample is 0.9717, which is similar to n = 0.5. Therefore, we maintain that the Hg heat reaction series of coal gangue during the process of combustion is 1.

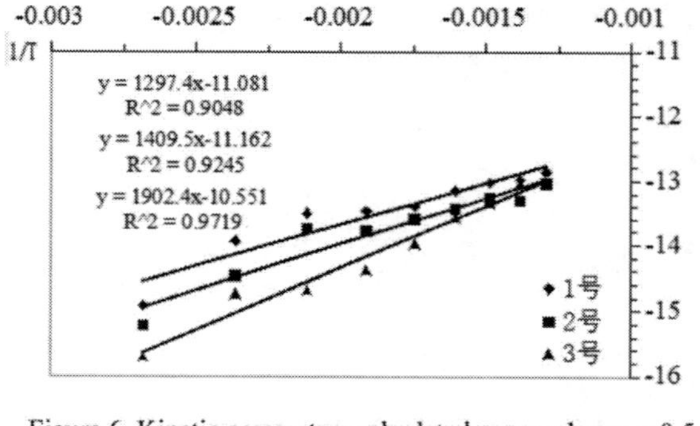

Figure 6. Kinetic parameters calculated curve when n = 0.5

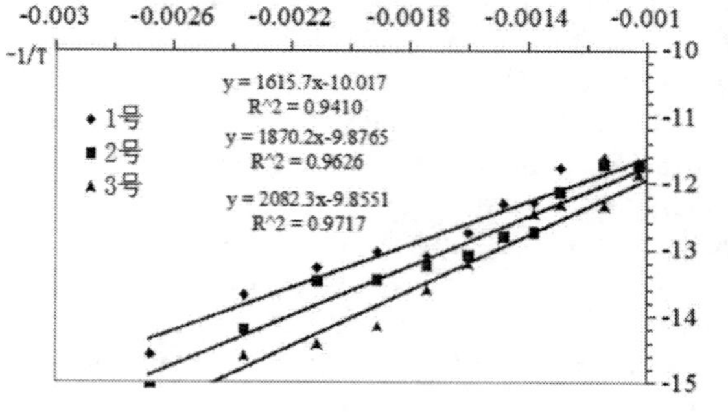

Figure 7. Kinetic parameters calculated curve when n = 1

After determining the value of n, we can determine the fitting equation of the straight line. Using the linear intercept and the slope, we calculated

E and *A*. The value of *n* in the process of the reaction series is assumed to be 1 for the NO. 3 samples. To determine the influence of activation energy and the former factor, the activation energy that refers to the former factor when $n = 0.5$ is considered to test if the hypothesis is reasonable. The calculation results of the reaction kinetics parameters of Hg in coal gangue when n = 0.5 and n = 1 are shown in Tables 3 and 4. The calculation results show that when n = 1, the activation energy of the three types of samples is similar. This assumption is reasonable.

Table 3. Calculation kinetic parameters of Hg in coal gangue when $n = 0.5$

$n = 0.5$	E (J/mol)	A (s^{-1})	R^2
1	10,233	0.02	0.92
2	12,050	0.03	0.928
3	15,541	0.06	0.979

Table 4. Calculation kinetic parameters of Hg in coal gangue when $n = 1$

$n = 1$	E (J/mol)	A (s^{-1})	R^2
1	14,550	0.10	0.9564
2	15,632	0.10	0.9504
3	17,760	0.13	0.9659

2.4 Hg emission characteristics in the burning process of coal gangue

2.4.1 Different fineness of the coal gangue combustion process and the effect of Hg release characteristics

Calculated separately for each coal gangue, the coal and air volumes and ash concentration in the samples measured in each condition are shown in Table 6.

Table 5. Mathematical expressions of common solid-phase thermal decomposition mechanism functions[15]

Number	Mechanism	Code name	G(α)	f(α)
1	One-dimensional diffusion	D1	α2	1/(2α)
2	One-dimensional diffusion	D2	α + (1 − α)ln(1 − α)	[−ln(1 − α)] − 1
3	One-dimensional diffusion (cylindrical symmetry)	D3	1 − 2α/3 − (1 − α)2/3	3/2[(1 − α) − 1/3 − 1] − 1
4, 5	Jander function (n = 2, 1/2)	D4	[1 − (1 − α)1/3]n	3/2(1 − α)2/3[1 − (1 − α)1/3] − 1
6	Jander function		[1 − (1 − α)1/2] 1/2	
7	One-dimensional diffusion	D5	[(1 + α)1/3 − 1]2	3/2(1 + α)2/3[(1 + α)1/3 − 1] − 1
8	One-dimensional diffusion	D6	[(1 − α) − 1/3 − 1]2	3/2(1 − α)4/3[1/(1 − α)1/3 − 1] − 1
9–16	Random successive nucleation growth (n = 1, 1.5, 2, 3, 4, 1/2, 1/3, 1/4)	An	[−ln(1 − α)]1/n	1/n(1 − α)[−ln(1 − α)](n − 1)/n
17	Contractile geometry (cylindrical symmetry)	R1	1 − (1 − α)1/2	2(1 − α)1/2
18	Contractile geometry (spherical symmetry)	R2	1 − (1 − α)1/3	3(1 − α)2/3
19			1 − (1 − α)1/4	
20			1 − (1 − α)2	
21			1 − (1 − α)3	
22			1 − (1 − α)4	
24–27	Index nucleation (n = 2/3, 1, 2, 3, 4)	Pn	α1/n	1/nα(n − 1)/n
28	Chemical reaction (level 2)	C2	(1 − α) − 1 − 1	(1 − α)2
29			(1 − α) − 1	1/2(1 − α)3
30	Chemical reaction (level 1.5)	C1.5	(1 − α) − 1/2	2(1 − α)3/2
31	Chemical reaction (level 3)	C3	(1 − α) − 2	1/2(1 − α)3

Table 6. Experimental data of samples with different fineness

	For coal (g/min)	To the air (L/h)	100 mesh sample Hg concentrations in ash (ng/g)	200 mesh sample Hg concentrations in ash (ng/g)	300 mesh sample Hg concentrations in ash (ng/g)
1	2.2	177	129.91	145.4	152.48
2	1	180	149.8	150.9	171.7
3	2.6	173	29.57	33.12	40.844

Figures 8 and 9 show the Hg content in the ash samples with different fineness and the percentage of Hg content in raw coal, respectively.

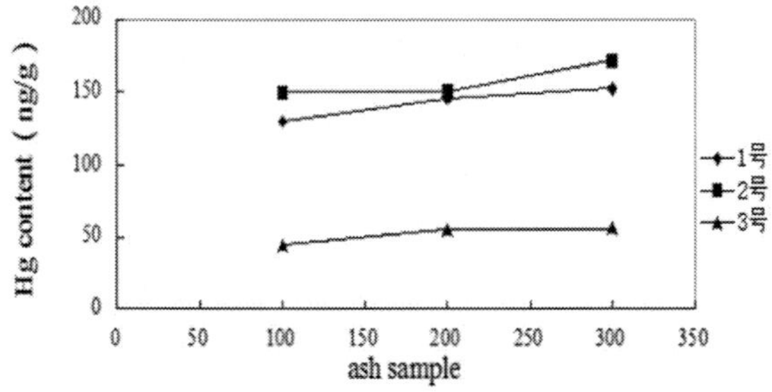

Figure 8. Different fineness of Hg content in the ash sample

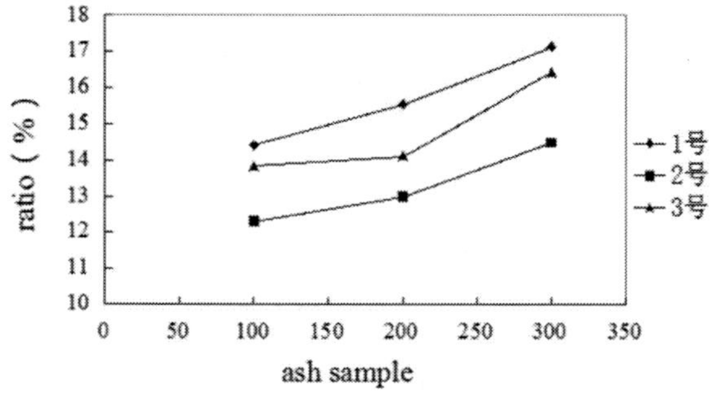

Figure 9. Samples with different fineness of Hg content in the ash of Hg content in the raw coal ratio

As shown in Figure 8, with the increase in sample fineness, the Hg content in ash after combustion in the three types of gangue is increased. The adsorption capacity of NO. 1 and NO. 2 coal gangue is approximately 150 ng/g, which is significantly different from that of the NO.3 coal gangue. The adsorption capacity of NO. 3 coal gangue is approximately 40 ng/g. The mercury concentration in ash was multiplied then divided by the mercury concentration in the raw coal to obtain the mercury content in the ash, which accounts for the proportion of the mercury content in the raw coal. Figure 9 shows that with the increase in pulverized coal fineness, the adsorption of mercury after burning in the ash increases. The adsorption of mercury in the three types of coal gangue samples is different from the proportion of adsorbed mercury in raw coal.

2.4.2 Different retention times and the effect of Hg emission characteristics on gangue combustion

Each coal gangue sample tested was picked from over 200 test samples based on the different fineness of the coal gangue combustion process and the impact of Hg release characteristics. Three groups were added as the additional test. The test data and the measured results are shown in Table 7.

Table 7. Experimental data of samples with different air volumes

	1			2			3		
Amount of coal (g/min)	1.6	2.2	2.8	1	1.6	2.2	1.2	1.9	2.6
Air volume (L/h)	130	178	230	180	290	400	80	126	170
Hg concentrations in the ash (ng/g)	161.6	144.4	108.4	150.9	115.7	70.13	63.14	57.87	49.35
Adsorption rate (%)	17.82	14.93	11.33	13.12	10.68	6.74	18.67	17.14	14.31

Figures 10 and 11 show the Hg content under different residence times in ash and the Hg adsorption ratio under different residence times in ash.

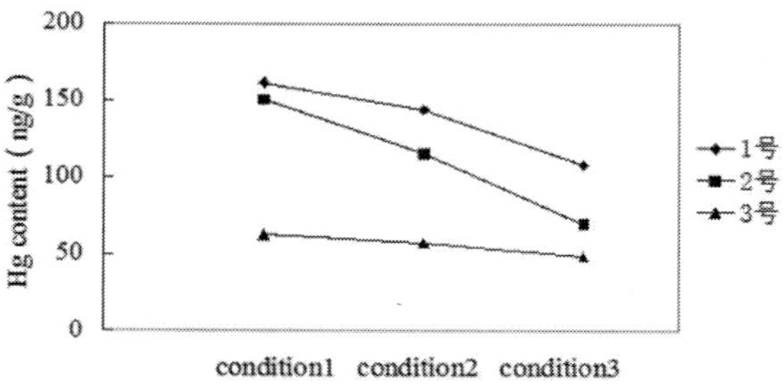

Figure 10. Hg content under different residence times in ash

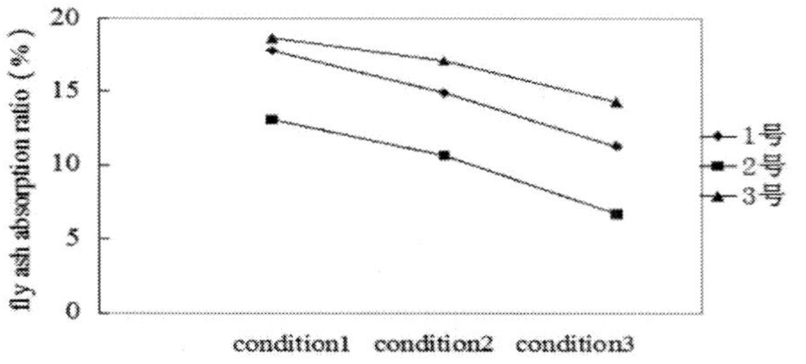

Figure 11. Fly ash adsorption ratio under different residence times in ash

As shown in Figures 10 and 11, with the increase in wind, the adsorption of Hg content in the three types of coal ash and the adsorption ratio decreased; the lower amplitude was enlarged. As to the increase in air flow, the flue gas flow rate increased and flue gas residence time in the furnace and cooling section decreased. The decrease in adsorption time may be the cause of the decrease. When flue gas residence time in the cooling section decreases, the flue gas flow rate and flue gas temperature both increase, resulting in a decrease in the fly ash adsorption amount of Hg. When wind is transferred, the Hg adsorption capacity and adsorption rate exhibit an obvious downward trend.

3 Conclusion

In this study, we analyzed the effect of temperature on coal gangue Hg emission characteristics[16–18], including the different heating terminal temperatures and the conditions of different heating times and heating rates as well as the effect of gangue Hg emission characteristics under different heating terminal temperatures based on the Hg reaction kinetics parameters. The Hg emission characteristics in the process of gangue combustion were studied. The following conclusions were obtained based on the results.

(1) Hg in coal gangue is stable at low temperature. At a final temperature of over 100 °C, the Hg content of the coal gangue samples decreased. The evaporation rate increased with the increase in temperature. At 500 °C, the Hg levels tend to be stable and the volatilization rate of Hg in coal gangue is >99%. The quality of release without Hg emission not only depends on the initial content of Hg in coal gangue but also on the geometrical shape of Hg.

(2) At the same temperature, the amount of Hg released in coal gangue increases gradually along with the increase in heating time. After a certain amount of time, the Hg release rate tends to stabilize. The Hg release rate decreases with time. For the same type of coal gangue but with a different temperature, the higher the temperature, the higher the release rate of Hg. The shorter the time to reach the steady state, the greater the amount of Hg released. At 300 °C, the Hg content in the coal gangue at 40 min remains constant. At 400 °C, the Hg content in the coal gangue requires 30 min to stabilize. The Hg levels at 500 °C for 10 min remain unchanged.

(3) The residue rate of Hg in coal gangue is not necessarily associated with the volatilization rate. The residual rate is determined by heating the end state of Hg in coal gangue. When the temperature is low, a fair amount of Hg is volatilized. In the coal pulverizing system and the coal feeding system in coal-fired power plants, at an air temperature of 200 °C or 300 °C, Hg volatilizes at a temperature rate that reaches 30% to 50%. When an additive or some other new technology is employed to release Hg, an ideal effect may be achieved.

(4) For the coal gangue heat reaction kinetics parameters of Hg, the Hg heat reaction series in the process of coal gangue is 1. The final results after solving are as follows: the heating Hg activation energies in the process of the three types of samples are 14,550, 15,632, and 17,760 J/mol corresponding to the former factors of 0.10, 0.10, and 0.13.

(5) The rate of Hg adsorption in coal gangue by adsorption of fly ash after burning is higher than that in the mixed coal combustion fly ash after burning, accounting for approximately 10% to 20%. This result is due to the high ash content of coal gangue. With the increase in sample fineness, the proportion of medium-sized and small-sized particles of fly ash, the fly ash specific surface area, and the adsorption capacity of Hg are increased.

(6) The fly ash and Hg concentrations in flue gas decrease along with the increase in wind; the flue gas flow rate increases. The cooling period requires a small amount of time, resulting in a smoke temperature increase. A few aspects of coal gangue fly ash decreased after combustion of Hg, such as the adsorption capacity, adsorption rate, and magnitude.

References

[1] Guo Yuqing. Comprehensive utilization of coal gangue countermeasures and meaning[J] Journal of Shanxi Coal-Mining Adminstrators College. 2011(1):43-45.

[2] Liu Zhiyi. Utilization status and problems of gangue resource[J]. Clean Coal Technology, 2011(1):105-107.

[3] Liang Tieshan, Gu Mingchuan. Phenomenon of overall increasing of temperature in fiery coal gangue heap and its theoretical explantion[J]. China Mining Magazine, 2009(4):93-95.

[4] Zhang Chengliang,Du Yongji,Li Meisheng,et al. Utilizing Heat Energy from Spontaneous Combustion of Coal Gangue Dump and its Utilization[J]. Journal of Anhui Agricultural Sciences.2008,36(6):2431-2432.

[5] Ren Jianli Zhou Jinsong, Luo Zhongyang, et al. Study of mercury emission during coal combustion[J]. ACTA SCIENTIAE CIRCUMSTANTIAE,2002,22(3):289-293.

[6] Yudovich Ya.E., Mercury in coal: a review Part 1. Geochemistry International Journal of Coal Geology,2005,62(3):107-134.

[7] Huang Wenhui, Yang Yichun. Mercury in Coal in China [J]. Coal Geology of China, 2002, 14(S1): 38-41.

[8] Tan Y, Mortazavi R, Dureau B, et al. An investigation of mercury distribution and speciation during coal combustion[J]. Fuel, 2004,83(16):2229-2236.

[9] Romero C E, Li Y, Bilirgen H, et al. Modification of boiler operating conditions for mercury emissions reductions in coal-fired utility boilers[J]. Fuel, 2006,85(2):204-212.

[10] M. Antonia Lopez-Anton, et al. Analysis of mercury species present during coal combustion by thermal desorption[J]. Fuel, 2010, 89(3):629-634.

[11] M.J.A natal,Jr G. Varhegyi. Cellulose pyrolysis kinetics: the current state of knowledge [J]. Ind.Eng.Chem.Res.1995, 34:703-717.

[12] Zhou Hua,Wang Qiang, ec al.Pyrolysis Characters and the Non-isothermal Kinetics of Shenhua Coal[J].Coal Chemical Industry，2010, 4(2)：27-31.

[13] H.N.Stiles,R.Kandiyoti.Secondary reactions of flash pyrolysis tars measured in a fluidized bed pyrolysis reactor with some novel design features [J]. Fuel, 1989, 68:275-282.

[14] Flynn J H. The 'Temperature Integral'—Its use and abuse[J]. Thermochimica Acta, 1997,300(1–2):83-92.

[15] HuRongzu,Shi Qizhen. Thermal analysis kinetics[M]. Beijing:Science & Technology Press,2001,2-4.

[16] Meng Lisu,Duan Yufeng,Huang Zhijun, ec al. Experimental Study on Factors Influencing Adsorption of Mercury Vapor by Coal-fired Fly Ash[J] Journal of Power Engineering, 2009,29(5):.487-491.

[17] Zhao Zhenguo. Adsorption principle of application[M]. Beijing: Chemical Industry Press,2005:63-75.

[18] Zhang YInzhi. Effect of porous structure and surface functionality on the mercury capacity of a fly ash carbon and it's activated sample[J]. Fuel, 2005,84(1): 105-108.

370720 Use of Ldpe Passive Samplers for Monitoring PAHs in the Water Column of a Suburban, Freshwater Oil-Spill Site

Abstract:

Polycyclic aromatic hydrocarbons (PAHs) are common components of petroleum products that are sometimes released to surface and ground water via leaks and spills, often threatening water quality and ecosystem health. Contaminants that dissolve into water are most readily available for biological uptake and exposure, and can be monitored using passive sampling techniques. This study used low-density polyethylene (LDPE) polymer sampling strips deployed for a period of four weeks in the riverine environment of a Grafton, MA canal that is heavily contaminated by Bunker-C heating oil. The PAHs that were sorbed by the LDPE from the aqueous phase were measured using gas chromatography-mass spectrometry (GC-MS). To permit calibration of the method, involving corrections for possible bio-fouling and differences in compound sampling rates, equilibrium dynamics were established using the loss of deuterated PAH reference compounds impregnated in the LDPE prior to deployment. Results using the LDPE samplers suggest that PAH concentrations previously determined by water grab-samples may overestimate the equilibrium water content of PAHs of concern. Results suggest the value of use of passive samplers in oil-spill contaminated sites, as an alternative to simple water grab samples.

Copyright © American Institute of Chemical Engineers. All rights reserved.

Residual solvent release from froth treatment tailings

Yuming Xu[1], Jianying Wu[1*], Tadek Dabros[1], Jianmin Kan[2]

1: CanmetENERGY, Natural Resources Canada, 1 Oil Patch Dr., Devon, Alberta T9G 1A8

2: Total E & P Canada, Suite 2900, 240 – 4th Avenue SW, Calgary, AB T2P 4H4

*: Current address, Canadian natural resources limited, 2500, 855 - 2 Street SW Calgary, Alberta, Canada T2P 4J8

Froth treatment tailings (TSRU) contain trace amounts of solvent. When they are deposited into tailings ponds under water or on a beach, solvent evaporates and is released to atmosphere. Local accumulations of solvent and high evaporation rates would raise safety and environmental concerns. It is therefore important to understand the behaviour of the solvent release from the TSRU tailings. Solvent release was studied using two approaches: 1) kinetic modeling, and 2) experiments simulating solvent release under two different scenarios: subaqueous deposition and subaerial deposition. A diffusion-controlled kinetic model was used to predict the solvent release from subaqueous deposition at temperatures below the solvent boiling point. A kinetic model based on continuity equations was found to successfully model solvent release from unsaturated tailings under aerial conditions at ambient temperature. However, at temperatures above the solvent boiling point, solvent release generates bubbles and produces foam. The foam formation changes the characteristic of tailings and makes the modeling very difficult. Experimental results demonstrated that solvent is released from subaqueous tailings very slowly due to the lack of agitation and mixing. Solvent trapped in the sediment did not effectively keep the overlying water saturated with pentane. Solvent release from aerially deposited beaches is faster than from subaqueous deposits; for instance, at 22°C after 24 h, only 15% of the solvent remained in aerially deposited tailings, compared to 90% remaining in subaqueous tailings under the experimental conditions.

380710 A Microscopic Analog of Saturated Soil for the Study of Colloid-Facilitated Transport of Contaminants

Abstract:

Although many contaminants have low solubility in water, they can adsorb on mobile colloids and travel with underground water flow over 80 m/year. Therefore, understanding colloid-facilitated transport is important for predicting the spatial and temporal distribution of contaminants. However, the detailed flow pathway and transport mechanism of those colloids are still elusive due to the conventional non-transparent experimental methods (*e.g.* soil cores) and the physical and chemical heterogeneity of the pore space. Here, we describe a microfluidic device to construct ananalog of saturated soil. Soft lithography was used to fabricate a device consisting of a rhombus chamber, inlet, and outlet channels in polydimethylsiloxane. The chamber dimensions are 1 mm for each side, 20 μm in height and 80° for the apex angle. The cross section of the inlet and outlet are 100μm×20μm and 1286μm×10μm, respectively. Due to the height difference between the chamber and outlet channel, ~4800 polystyrene spheres (15 μm) weretrapped and packed one-by-one to form the porous medium within the chamber.The physical and chemical heterogeneity in the bead-packed arrays can be decoupledby using different sized beads with varying surface properties. The linear velocity of flow at inlet was 1μm/s to 25 mm/s, yielding Peclet numbers of that range from 10 to 10^5. We also used the Lattice Boltzmann (LB) method to simulate the flow field and colloidal transport in the media. The computational domain was faithfully re-constructed from the experimental porous media using a custom image processing routine. We used the Canny algorithm to detect the outline of each bead and obtained its x-y coordinate based on the circularity of each outline. The z-position was estimated by adjusting focal plane to achieve the sharpest contrast of the bead edge. Using this approach, we measured and simulated colloidal transport trajectories and extracted key parameters such as the breakthrough curve, effective retention, and entrapped colloids distribution. We will discuss the comparison between our experiments and LB simulations results, which reveal the rate-limiting transport mechanismunder different flow conditions.

Copyright © American Institute of Chemical Engineers. All rights reserved.

368725 Micro and Macro-Heterogeneities Effects on Transport of Pollutants in Porous Media: Fundamental Modeling Aspects

Abstract:

Soils are very complex porous media with a variety of scales playing a role on solutes and contaminants transport. Complete fundamental understanding about transport in complex porous media, such as soils, is far to be complete. Experimental approaches when an electrical-field-based technique is applied in homogeneous porous media are commonly reported in the literature. However, there are few modeling studies on this field to predict the role of multiple scales and driving forces on the soil-cleaning efficiency. Regarding heterogeneous porous media, where heterogeneities at *the micro and macro-scale* are included are in general difficult to find in the literature. Based on this motivation, this study attends to develop a more effective, systematic, and predictive description of transport of contaminants under removal operations for soils. Here, the porous structure of a soil is described by idealized micro-capillaries of a single dimension; which change both their orientation and geometry in the soil matrix. Therefore, Micro-heterogeneities will be analyzed by switching two micro-capillary geometries and assessing their impact on contaminant transport. By using an Area Averaging Approach the transport of a contaminant in micro-capillaries is scaled-up to a macro-transport equation within the capillary scale where the geometry changes will be assessed. In addition, the role of macro-heterogeneities is described by including gravity effects that depends on the capillaries inclination angle. Preliminary results show effective parameters varying according to micro and macro-heterogeneities as well as operating conditions. These results show an important effect of macro-heterogeneities on the effective parameters and averaged species concentration profile. Important outcomes from this study include, for example, cleaning soil times, and relevance of porous media structure; useful for technology developments such as site selection for waste disposal, and geo-membranes design.

Copyright © American Institute of Chemical Engineers. All rights reserved.

Experimental Study The Influence of Bromine、Iron on Particulate Mercury Formation

Qisheng Xu[1], Peixin Wu[2], Shaokun Lv[2], Zhengyang Gao[2]
1 Electric Power Research Institute of Guangdong Power Grid Corporation, Guangzhuo, 510000, Guangdong Province, China
2 School of Energy and Power Engineering, North China Electric Power University, Baoding 071003, Hebei Province, China

Abstract: Br、Fe has a significant impact on particular Hg (Hg_p) formation in the flue gas. This paper conducted burning experiments for two kinds of coals blending with KBr, $CaBr_2$, Fe_2O_3 and KBr with Fe_2O_3 together in the high-temperature electrically heated reactor. The addition ratio of KBr, $CaBr_2$ was 0.1%、0.3%、0.5%, combustion temperature was 1250℃. The fly ash samples of different combustion condition were collected, and the Hg content in fly ash was measured. The result shows that: the Hg content in ash and particulate Hg ratio increase with increased KBr, $CaBr_2$ addition ratio. At the same addition ratio, adding $CaBr_2$ led to slight higher Hg_p formation than adding KBr. The influences of Fe_2O_3 addition on Hg_p fromation are different for the two coals. The generation of Hg_p for coal 1 increases linearly with the increase of Fe_2O_3, however the generation of Hg_p for coal 2 has no significant promotion with Fe_2O_3 addition. But at KBr additions, Hg_p formation is also promoted by Fe_2O_3 addition for coal 2.

Key words: Br; mercury; Fe_2O_3; coal

0 INTRODUCTION

China's power is domanited by coal-fired power generation. The mercury exists in raw coal, which will be released in the combustion process to pollute the environment. At present, controling of mercury emissions from coal-fired process has become a hot topic in the research of [1-3] at home and abroad. China's 《Emission standard of thermal power plant's air pollutants》(GB 13223-2011) defined that the Hg concentration of flue gas that emissioned by the new coal-fired boiler is lower than $0.03mg/m^3$. Therefore, it is necessary to discuss the reactive characteristic of Hg and the reaction mechanism of Hg in flue gas, which is statistically significant to control mercury emission of coal-fired boiler.

Hg of flue gas in the boiler tail heating surface exists in the three forms of gaseous elemental Hg (Hg^0), gaseous two price Hg (Hg^{2+}) and particulate Hg (Hg_p). Hg^0 does not dissolve in water and is not easy to be adsorbed by the adsorbent, which will pollute the environment after emissions. Hg^{2+} can dissolve in water and can be removaled by wet desulphurization equipment, but also is easy to be adsorbed by the adsorbent, while in wet desulfurization process, part of the Hg^{2+} may be reduced to Hg^0 formation and be released, the adsorption with adsorbent is high cost; the Hg_P can be trapped by dust removal equipment[4,5]. If Hg^0 can be promoted to more selectively transform to Hg_p, it can realize the purpose of reducing mercury emission economically. But at present, the specialized study on promoting Hg_p formation are still relatively small.

A lot of researchs on Hg reaction of flue gas indicate that Cl plays a key role in the oxidation process of elemental mercury and the formation of Hg_P, Br and Cl are the same kind of halogen, Br has the similiar effect on the oxidation process of Hg^0[6,7] , but in the flue gas temperature, substances which contain Br decompose more easily and produce Br atom and Br_2 to react with the Hg in flue gas. In addition, Br can also react with HCl to promote the reaction of Cl and Hg, therefore, the amount of Br needs to be far less than Cl to achieve the same level oxidation and granulation of Hg. After addition of Br, there is not the formation of a large number of HBr in the flue gas to lead to the problem of corrosion on the heating surface. Although, for the promotion of oxidation and granulation of Hg^0, Br is better than Cl. However, there is still a lack of theoretical analysis and experimental study on what kind of containing Br substance can more effectively promote the conversion of Hg^0 to Hg_p, whether the effect of Hg_p generation is the same for different kinds of coal added with the same containing Br substances, and how much should be added.

In addition, some metal oxides can directly oxidize Hg^0 or catalyze the oxidation of Hg^0 through the catalytic formation of Cl_2, Fe_2O^3 is a common metallic oxides in the coal ash. But different scholars give different results with whether Fe_2O_3 can promote Hg^0 oxidation, whether addation of Fe_2O_3 and Br will produce a synergistic effect to the more effectively promote the oxidation of Hg^0 is also needed to fully study.

Therefore, two experiments are taken in the high temperature electric heating reactor with two kinds of coal added with KBr, $CaBr_2$, Fe_2O_3 and adding KBr and Fe_2O_3, fly ash of the combustion process of the acquisition, Several fly ash samples are collected and the Hg content is analysed, according to the analysis results, this paper discusses the formation rule of Hg_p under different combustion conditions. The results showed that two kinds of coal added with KBr, $CaBr_2$ can promote the formation of Hg_p. The formation rule of Hg_p after the addition of Fe_2O_3 is different in two kinds of coal, but on the basis of KBr ,two kinds of coal added with Fe_2O_3 can play a promoting effect on the formation of Hg_p.

1 TEST METHOD AND CONDITION

The high temperature electric heating reactor is shown in figure 1. The heating section is alundum tube, which can prevent the mercury from being adsorbed on tube wall with the decrease of gastemperature. The combustion temperature is 1250℃ and the gastemperature of the quartz tube section is measured along the flow direction. Then, place the sootcollector in the point where the temperature is 150℃ to collect fly ash sample. The feeding quantity is 2g/min and the excess air factor is 1.3.Hg content in coal and ash samples use RA-915M analyzer of Lumex company to measure, measurement error of the instrument is less than 10%.

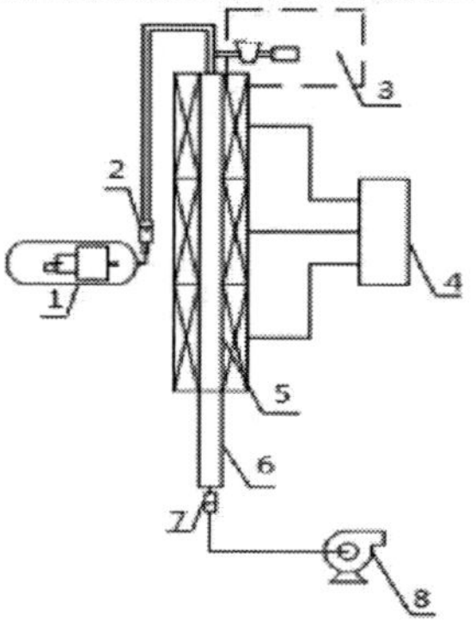

Fig. 1 The combustion experiment system

1—air compressor; 2—flow meter; 3—feeder; 4—temperature-controlled cabinet ; 5—furnace (alundum tube, insulating layer); 6—quartz tube (an extension of the alundum tube used as yhe cooling section of flue gas); 7—sootcollector; 8—induced draft fan

The industrial and elemental analysis results are shown in table 1. The Hg content of No.1 and No.2 coal is 207ng/g ,118 ng/g, respectively. The content of Cl of No. 1 coal is 3.6 times of No.2 coal, and the content of ash is more than 2 times No.2 coal. For the actual power plant boiler, if the content of Br added in the coal is too large, it may cause the heating surface corrosion, but also the cost is very high, the amount of Br should also be same with the level of Cl content in coal, but considering that residence time of the Flue Gas in the experimental furnace is limited(generally residence time of flue gas in power plant boiler tail heating surface is above 5S, but residence time was about 1.2s during experimental furnace combustion period, and residence time is about 0.4s during cooling period), the mixing effect of all components in the flue gas is worse than the actual boiler, with Cl content of two species of coal, the added amount of KBr, $CaBr_2$ is set to 0.1%, 0.3%, 0.5%, it is an order of magnitude higher than Cl content of two kinds of coal. Because Fe_2O_3 mainly plays a catalytic role, with the addition of KBr, $CaBr_2$,the amount of Fe_2O_3 is also set to 0.1%, 0.3%, 0.5%.On the basis of the above test, Also add 0.1%, 0.3%, 0.5% KBr, while adding 0.3%, 0.8% Fe_2O_3 combustion test. It can be seen that the added Fe is less than coal containing Fe.

Table 1 Properties of the coals

coal	Proximate analysis $_{ad}$ /%			Ultimate analysis $_{ad}$ /%							
	M_{ad}	A_{ad}	V_{ad}	C_{ad}	H_{ad}	O_{ad}	N_{ad}	S_{ad}	Cl	Ca	Fe
1	1.51	27.02	11.18	64.26	2.95	1.58	1.01	1.67	0.036	0.39	1.52
2	5.83	11.6	28.34	70.83	4.34	5.51	1.47	0.42	0.01	1.15	0.78

2 TEST RESULT AND ANALYSIS

2.1 Burnout characteristics of coal in differen conditions

Many studies suggest that the C catalyzes the oxidation of Hg^0 in the fly ash and granulated[8,9].To exclude the influence of the C-containing fly ash as much as possible, in order to clear the Br add Hg_p generated studied the carbon content of the fly ash is collected for each test procedure for measured. The result of the measurement shows that, At the time of this test two kinds of coal raw and add $CaBr_2$, KBr, of Fe_2O_3, ash carbon and not with the changes in the proportion of each substance is added exhibit regularity trend. The conditions of the 1st coal fly ash containing C change in the range of 3.4% -4.6%, the conditions of the 2nd coal fly ash containing C change in the range of 4.2% -5.8%. Two coal fly ash containing the C of different reasons has two main aspects, First, the characteristics of the combustion of the combustible components of the two coals, the second of the two coal ash content. The 1st ash content is 27.02%, much higher than 11.6% of the coal, obviously, coal fly ash containing C values appear below the 2nd coal, due to its high ash content, the degree of the 1st coal burnout than the 2nd coal.

For the fly ash containing C generated on the Hg_p, the subsequent analysis showed that the two coal add $CaBr_2$ KBr, Fe_2O_3, fly ash Hg content added with each component of the proportion of the change regularity trend showed, Contrast fly ash containing C does not add with each component exhibit regularity trends can be identified, in this experiment, fly ash containing the C content generated particles Hg far less than $CaBr_2$, KBr, Fe_2O_3 add impact.

2.2 Generative rule of Hg_p with KBr addition

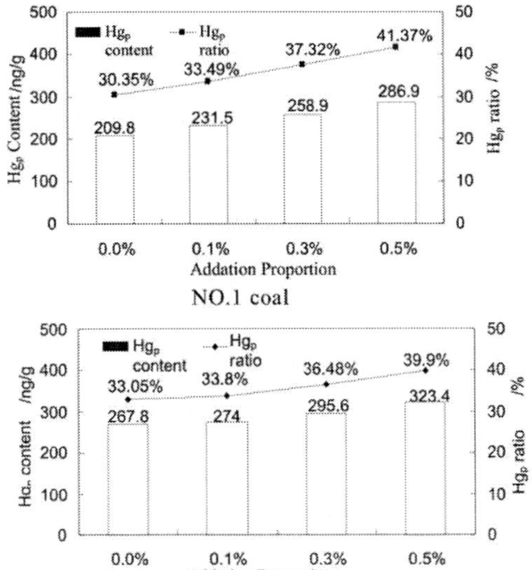

NO.2 coal

Figure. 2 Hg_p contents and Hg_p fractions of the two coals with KBr addition

The Hg_P content and fraction of the coal with KBr addition is shown as figure 2. Hg_p content is the Hg concentration in the fly ash, Hg_p fraction is Hg_p accounted the total Hg percentage of carry into the furnace to the coal. As can be seen from the diagram data, two coal combustion fly ash Hg content similar and Hg_p ratio almost the same, both slightly higher than 30%. Add KBr showed an upward trend with the increase in the amount of Hg_p add content and fraction. Hg_p content and fractional increase of KBr in increases, the Br added with KBr to play a catalytic role in the formation of particles of Hg.

Flue gas, Hg^{2+} ash particles adsorbed Hg^0 easier, so Promote Hg^0 oxidation is promoted particle Hg_p generation. Further, in the flue gas environment, with the reactive groups and flue gas in correlation components between non-homogeneous reaction, and the different stages of the reactants with the reaction product on the ash particles of adsorption and desorption, Hg^0 oxidation overlap associated with cross with granulation. Further, in the flue gas environment, with the reactive groups on the ash particles and flue gas correlation components between non-homogeneous reaction, and the different stages of the reactants with the reaction product on the ash particles of adsorption and desorption, Hg0 oxidationoverlap associated with cross with granulation. The data in Table 1 show that the two coals contain certain Cl, Cl and Br are the elements of the same family, both of reaction with Hg similar, for homogeneous reaction. When Coal combustion, Cl in the coal all released in the form of HCl in the flue gas. With the decline temperature of the flue gas, HCl can be decomposed in accordance with the following reaction Cl atoms and Cl_2:

$$HCl+OH \rightarrow Cl+H_2O \quad (1)$$

$$Cl+Cl \rightarrow Cl_2 \quad (2)$$

HCl, Cl, Cl_2 all can react with Hg, the overall reactions are expressed as :

$$Hg+Cl_2 \rightarrow HgCl_2 \quad (3)$$

$$Hg+2HCl \rightarrow HgCl_2+H_2$$

(4)

HCl and Hg direct reaction barrier largest and reaction hardest. The most vulnerable reaction of the Cl atom and Hg is carried out at a lower temperature. Cl atoms generated significant need higher temperatures, which makes small chance by the Cl atom oxidation Hg. Some metal oxides can promote HCl decomposition Cl_2, Therefore, the reaction (3) is important method of heterogeneous oxidation of Hg in the flue gas. Br and Hg reaction is similar, While Br reaction with HCl, and thus plays a role in promoting Cl oxidation of Hg. Specific overall reactions are expressed as:

$$Hg+Br_2 \rightarrow HgBr_2 \quad (4)$$

$$4HBr+O_2 \rightarrow 2Br_2+2H_2O \quad (5)$$

$$HBr+HCl+1/2O_2 \rightarrow BrCl+H_2O \quad (6)$$

$$Br_2+HCl \rightarrow BrCl+HBr \quad (7)$$

$$2BrCl \rightarrow Br_2+Cl_2 \quad (8)$$

$$2Hg+2BrCl \rightarrow HgBr_2+HgCl_2 \quad (9)$$

According to the reaction (7), (8), Br promote HCl decomposition. Reaction (10) and Hg reaction may generate BrCl, which is in accordance with the decomposition of the reaction (9) and then in accordance with reaction (3), (5) is reacted with Hg^0. Obviously adding Br can promote Hg^0 oxidation and ash particles are easily absorbed to the Hg^{2+}, which provides a mechanism to explain that the particles Hg_p ratio increase after added Br. Addition to the homogeneous reaction pathway, Cl, Br, and other elements can also be the effect of the active sites on the ash particles of C, Ca and other elements through non opposite response Hg is oxidized, and contributed to the formation of particles of Hg. The reaction mechanism is expressed as:

$$A(g) \rightarrow A(ads) \quad (11)$$

$$B(g) \rightarrow B(ads) \quad (12)$$

$$A(ads)+B(g) \rightarrow AB(ads) \quad (13)$$

$$A(ads)+B(ads) \rightarrow AB(ads) \quad (14)$$

$$AB(ads) \rightarrow AB(g) \quad (15)$$

In The above reaction formula, (g) indicates that the substance is gaseous,(ADS) indicates that the substance is adsorbed on the particles. Reaction (11) - (13) are as a Hg^0 the heterogeneous oxidation mechanism given. However, if we use A represents Hg^0, the reaction product of reaction (11), (13), (14) are Hgp. It also shows that Hg on the ash particles may not become particles Hg, one hand the final chemical speciation of these Hg is depends on the adsorption key states and key energy on the ash particles, also it's related to whether it has further reaction. Moreover, Cl or Br in the flue gas can be passed through the reaction (13) with some Hg have been attached in the ash particles, which is already into particles Hg. first of all, combined with these Hg, and then these Hg peeled from the ash particles by reaction (15) to become gaseous Hg. Visible Cl or Br can not only through some ways promote Hg_p generated, but also can inhibit Hg_p generated through some other ways. The generation ratio of Hg_p depends on the reaction of these two aspects competition. Except the role of Br, the K that KBr carrying will affect the oxidation of Hg. The study of the literature [13] shows that the presence of SO_2 from flue gas to the inhibition of the oxidation of mercury, but the alkali metal mainly by the interaction between the Cl ,S to weak the influence of the SO_2 so that play a role to promote Hg oxidation. We can be seen from figure 2, when the KBr additive level is 0.1%, even if the amount of Br with KBr added is much higher than the Cl of the raw coal. It means that although the halogen content has increased substantially with the feed powder into the reactor, the amount of particles Hg did not show a substantial increase. An important reason leading to this phenomenon is the limited blend conditions, the residence time and the reaction of Hg proceed sufficiently. It can be speculated that the actual coal-fired boiler with the equivalent bromine salt effect will be more evident than this test because of the respective components in the flue gas can be sufficiently blended and the flue gas residence time is longer.

2.3. Generative rule of Hg_P with $CaBr_2$ addition

The Hg_P content and fraction of the coal with KBr addition is shown as figure 3. As can be seen from the data in the graph, with $CaBr_2$ added the amount of Hg_p in the ash content and the ratio is also increasing, apparently adding $CaBr_2$ can promote the generation of particles Hg. Compare Figure 3 with Figure 2 it can be seen that the formation of particles of Hg in adding $CaBr_2$ is slightly higher than that of adding KBr conditions in the same add mass ratio.

In the combustion of pulverized coal, $CaBr_2$ reacts with H_2O in the flue gas to form CaO, $Ca(OH)_2$ and HBr(reaction 16,17). $Ca(OH)_2$ decomposes into CaO(reaction 18). Ca can directly provide the adsorption of Hg in the flue gas with active points, and CaO can directly adsorb $HgCl_2$[14,15] while the efficiency of adsorption of $HgCl_2$ by $Ca(OH)_2$ is higher. Visibly, Hg_p formation can be promoted by Br in $CaBr_2$, and Ca in $CaBr_2$ also can have the same effect. So adding $CaBr_2$ leads to higher Hg_p formation than adding KBr.

$$CaBr_2+H_2O \rightarrow CaO+2HBr \quad (16)$$

$$CaBr_2+2H_2O \rightarrow Ca(OH)_2+2HBr \quad (17)$$

$$Ca(OH)_2 \rightarrow CaO+H_2O$$

(18)

However, it is worth noting that the usage of these two kinds of salts content is controlled by mass ratio in the experiment. The molecular weight of KBr is smaller than $CaBr_2$, but a KBr molecule contains only one Br atom while a $CaBr_2$ molecule contains two Br atoms. So the amount of Br contained by these two salts is different at the same addition mass of the two salts. At the same mass, the amount of Br in KBr is 84% of the amount in $CaBr_2$. In addition, Ca can also adsorb Hg gas to form particular Hg while K in KBr has a vague impact on Hg in the flue gas. Comparing figure 2 with figure 3, it can be seen that particular Hg formation ratio with KBr addition is more than 95% of the ratio with $CaBr_2$ addition at the same addition mass. Just from the view of Br addition, adding KBr has a more significant effect on particular Hg formation under the experimental conditions in this paper.

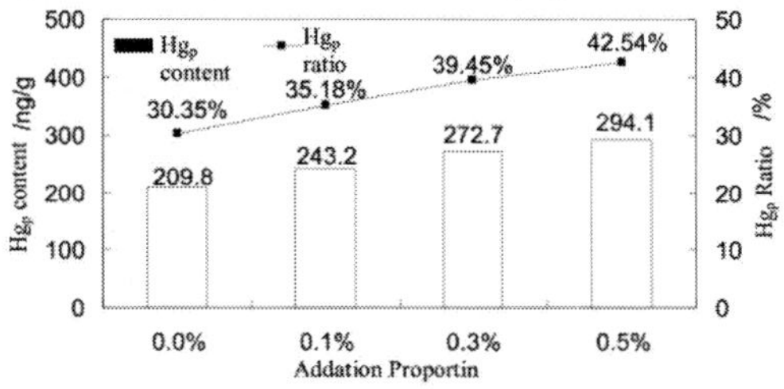

(a) coal 1

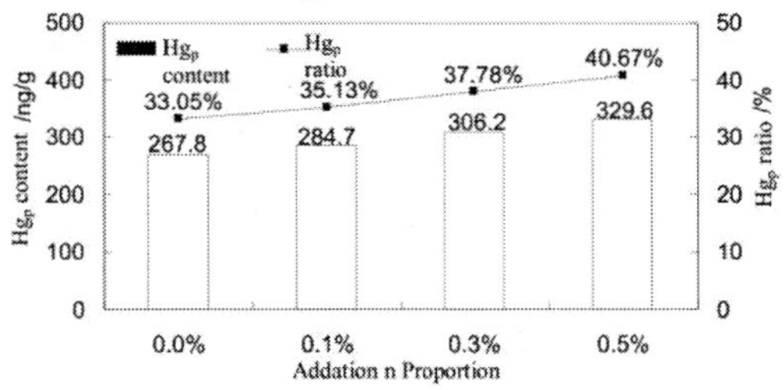

(b) coal 2

Figure. 3 Hg_p contents and Hg_p fractions of the two coals with $CaBr_2$ addition

2.4 Hg_p formation characteristics with Fe_2O_3 addition

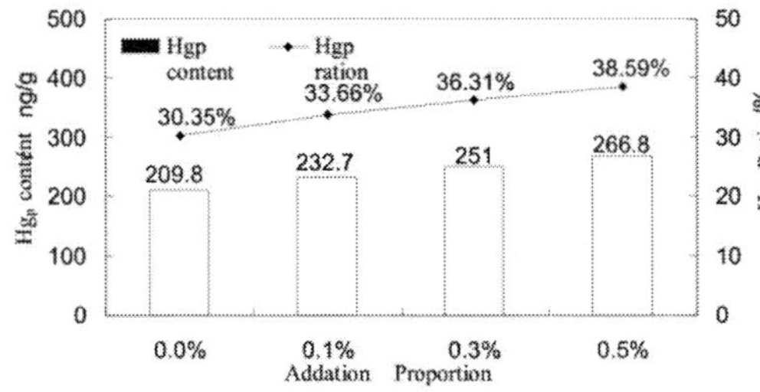

(a) coal 1

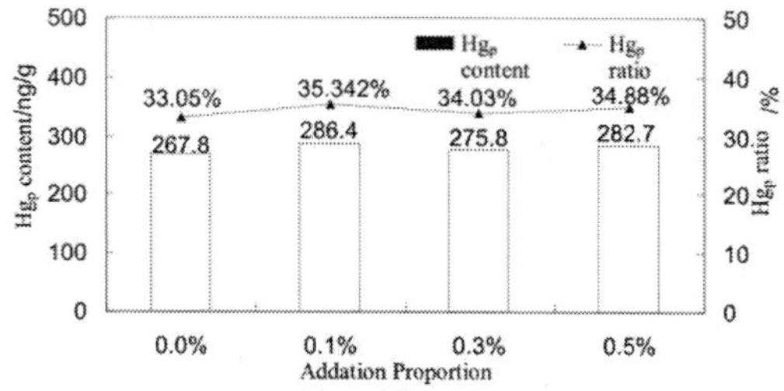

(b) coal 2

Figure. 4 Hg_p contents and Hg_p fractions of the two coals with Fe_2O_3addition

From the foregoing reaction mechanism, it can be seen that the reactions between Hg^0 and Cl_2, Br_2 which respectively generated from the decomposition of HCl and HBr are critical to the oxidation of Hg^0. Under the atmosphere of flue gas and the temperature level when coal is burnt, metal oxides can promote the decomposition of HCl and HBr. Fe_2O_3 is one kind of metal oxide. It remains to be further studied whether Fe_2O_3 addition could promote the formation of particulate Hg or not. Fe_2O_3 mainly exists in two forms–α-Fe_2O_3 and γ-Fe_2O_3. γ-Fe_2O_3 is metastable under natural conditions. Take it into account that the majority of Fe_2O_3 in nature is α-Fe_2O_3, α-Fe_2O_3 was selected in this experiment although γ-Fe_2O_3 addition can promote the oxidation of Hg in the flue gas[16].

Figure 4 are the particular Hg formation characteristics of burning experiments for two kinds of coals blending with Fe_2O_3. It can be seen that the generation of Hg_p for coal 1 increases with the increase of Fe_2O_3, and in the range of the addition in this paper, the generation of Hg_p increases linearly with Fe_2O_3 addition. So Fe_2O_3 can promote the oxidation and graining of Hg^0. This conclusion is consistent with the literature[17,18]. However, the generation of Hg_p for coal 2 has no significant promotion with the Fe_2O_3 addition. This is consistent with the literature[16]. After analysis, it is convinced that these results are caused by the different amount of Cl contained by these two coals. Data in Table 1 indicate that Cl content of coal 1 is 3 times more than coal 2. Cl content of coal 2 is low, so the HCl content in the flue gas for coal 2 is lower than coal 1. Adding Fe_2O_3 for coal 2 couldn't play a effective role in catalyzing the reaction between Hg and HCl. Figure 5 are Hg_p fractions when burning the two kinds of coals blending with 0.3%, 0.8% Fe_2O_3 together with 0.1%, 0.3%, 0.5% KBr. It can be seen from Figure 5.a and Figure 5.b that at any addition ratio of KBr, the oxidation and graining of Hg^0 can be promoted by Fe_2O_3 addition, whether for coal 1 or coal 2. Hg_p ratio when adding 0.8% Fe_2O_3 is slight higher than adding 0.3% Fe_2O_3. Visibly, under the condition of plenty of halogen, Fe_2O_3 can promote Hg^0 oxidation and graining. It also suggests that Fe_2O_3 couldn't directly absorb Hg in the flue gas or react with Hg. Fe_2O_3 can promote the oxidation and graining of Hg in the flue gas during promoting the reaction between halogen and Hg.

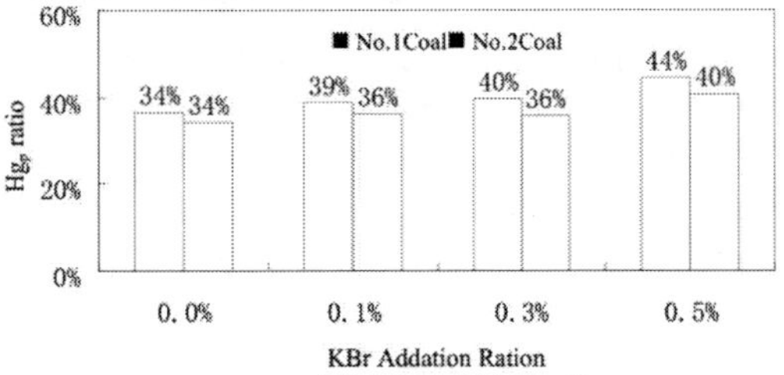

(a) Add 0.3% Fe_2O_3

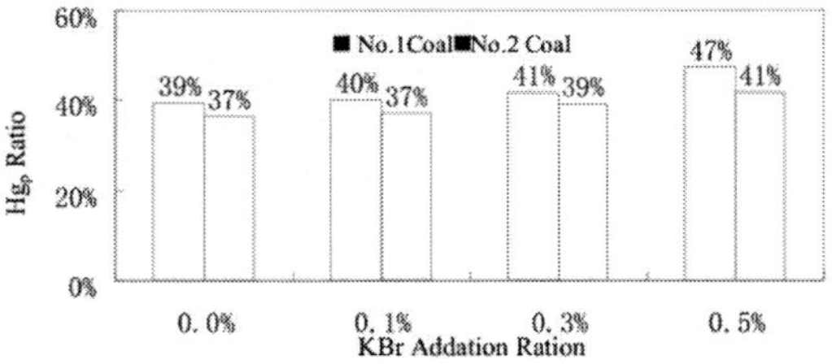

(b) Add 0.8% Fe_2O_3

Figure. 5 Hg_p fractions of the two coals with Fe_2O_3 and KBr addition

3 CONCLUSION

According to the experimental results and analysis, the following conclusion can be drawn:

(1) The Hg content in ash and particulate Hg ratio increase with increased KBr, $CaBr_2$ addition ratio. At the same addition adding $CaBr_2$ leads to slight higher Hg_p formation than adding KBr.

(2) At the same mass, the amount of Br in KBr is 84% of the amount in $CaBr_2$.In addition, Ca can also adsorb Hg gas to form particular Hg while K in KBr has a vague impact on Hg in the flue gas. From the view of Br addition, adding KBr has a more significant effect on particular Hg formation under the experimental conditions in this paper.

(3) in the range of the addition in this paper, the generation of Hg_p for coal 1 increases linearly with the increase of Fe_2O_3, however the generation of Hg_p for coal 2 has no significant promotion with Fe_2O_3 addition.

(4) Fe_2O_3 couldn't directly absorb Hg in the flue gas or react with Hg. Fe_2O_3 can promote the oxidation and graining of Hg in the flue gas during promoting the reaction between halogen and Hg.

Reference

[1] M. Antonia López-Antón, Mercedes Díaz-Somoano, Raquel Ochoa-González, M. Rosa Martínez-Tarazona. Analytical methods for mercury analysis in coal and coal combustion by-products[J]. *International Journal of Coal Geology*, 2012, 94(1): 44-53.

[2] Jennifer Wilcox, Erik Rupp, Samantha C. Ying, Dong-Hee Lim, Ana Suarez Negreira, Abby Kirchofer, Feng Feng, Kyoungjin Lee. Mercury adsorption and oxidation in coal combustion and gasification processes[J]. *International Journal of Coal Geology*, 2012, 90-91(1): 4-20.

[3] Liugen Zheng, Guijian Liu, Chen-Lin Chou. Abundance and modes of occurrence of mercury in some low-sulfur coals from China[J]. *International Journal of Coal Geology*, 2008, 73(1):19-26.

[4] Deepak Pudasainee, Jeong-Hun Kim, Young-Sik Yoon, Yong-Chil Seo. Oxidation, reemission and mass distribution of mercury in bituminous coal fired power plants with SCR, CS ESP and wet FGD[J]. *Fuel*, 2012, 93: 312-318.

[5] Ye Zhuang, Jeffrey S. Thompson, Chris J. Zygarlicke, John H. Pavlish. Impact of calcium chloride addition on mercury transformations and control in coal flue gas[J]. *Fuel*, 2007,86(15): 2351-2359.

[6] Shou-Heng Liu, Nai-Qiang Yan, Zhao-Rong Liu, Zan Qu, H. Paul Wang, Shih-Ger Chang, Charles Miller. Using Bromine Gas to Enhance Mercury Removal from Flue Gas of Coal-Fired Power Plants[J]. *Environmental Science and Technology*, 2007, 41(4): 1405-1412.

[7] Kevin J. Hughes, Lin Ma, Richard T.J. Porter, Mohamed Pourkashanian. Mercury Transformation Modelling with Bromine Addition in Coal Derived Flue Gases[J]. *Computer Aided Chemical Engineering*, 2011, 29: 171-175.

[8] P. Abad-Valle, M.A. Lopez-Anton, M. Diaz-Somoano, M.R. Martinez-Tarazona. The role of unburned carbon concentrates from fly ashes in the oxidation and retention of mercury[J]. *Chemical Engineering Journal*, 2011, 174(1): 86-92.

[9] M. Antonia López-Antón, Patricia Abad-Valle, Mercedes Díaz-Somoano, Isabel Suárez-Ruiz,M. Rosa Martínez-Tarazona. The influence of carbon particle type in fly ashes on mercury adsorption[J]. *Fuel*, 2009, 88(7): 1194-1200.

[10] N.V. Manukyan, V.H. Martirosyan. Investigation of the chlorination mechanism of metal oxides by chlorine[J]. *Journal of Materials Processing Technology*, 2003, 142(1): 145-151.

[11] Sliger R., Kramlich J., et al. Towards the development of a chemical kinetic model for the homogeneous oxidation of mercury by chlorine species.[J] *Fuel Processing Technology*, 2000, 65():423-438

[12] Bjoern Hall, Oliver Lindqvist, Evert Ljungstroem. Mercury chemistry in simulated flue gases related to waste incineration conditions[J]. *Environmental Science and Technology*, 1990,24(1): 108-111.

[13]Yu Wanxuan, Diao Yongfa, Shen Heng. Thermodynamics study of effects of alkali metals on mercury transformation during co-combustion of biomass with coal[J]. *Proceedings of the CSEE*, 2012，32(26):96-101(In Chinese).

[14] Mahamud Subir, Parisa A Ariya, Ashu P. Dastoor. A review of uncertainties in atmospheric modeling of mercury chemistry I. Uncertainties in existing kinetic parameters-Fundamental limitations and the importance of heterogeneous chemistry[J]. *Atmospheric Evironment*, 2011, 45(32):5664-5676.

[15] Guo Xin, Pengfei Zhao, Chuguang Zheng. Theoretical study of different speciation of mercury adsorption on CaO (001) surface[J]. *Proceedings of the Combustion Institute*, 2009, 32(2): 2693-2699.

[16] Kevin C. Galbreath, Christopher J. Zygarlicke, James E. Tibbetts, Richard L. Schulz, Grant E. Dunham. Effects of NOx, α-Fe_2O_3, γ-Fe_2O_3, and HCl on mercury transformations in a 7-kW coal combustion system[J]. *Fuel Processing Technology*, 2004, 86(4):429-448.

[17]Huawei Huang，Jingjin Luo. Effect of Various Fly Ash Compositions on Mercury Speciation Transformation [J]. Proceedings of the Chinese Society for Electrical Engineering，2010，30：70-75.

[18]Mercedes Diaz-Somoano，Sven Unterbei，Klaus R G Hein，Using wet-FGD systems for mercury removal[J] . Journal of Environmental Monitoring,2005,7(9):906-909.

363681 Impacts of Particle-Based Dispersants on Benzene Bioavailability and on Toxicity Using Artemia Franciscana As a Model Organism

Abstract:

Nanoparticles are able to assemble at water-oil interfaces and stabilize fine oil droplets, forming what is known as a Pickering emulsion. Nanoparticles are a possible alternative to the chemical dispersants typically deployed after an oil spill. Both particle and chemical based dispersants serve to limit petroleum accumulation at the ocean surface and accelerate biodegradation processes by increasing the oil-water interfacial area. The potential advantage of our carbon black particle –based Pickering emulsion is the ability to not only stabilize droplets but also to adsorb the more water-soluble and toxic fractions of petroleum. Benzene is used as a model aromatic compound to determine the efficacy of the carbon black particles to adsorb and retain (therefore reducing the bioavailability) these fractions as a function of tunable hydrophilicity. *Artemia franciscana* (brine shrimp) larvae are used as a model marine microcrustacean to assess toxicity. Endpoints for the larvae include activation of stress responses, oxidative damage, reduced motility and death. This study provides a novel, interdisciplinary approach for rational design of surface-engineered nanoparticle-based dispersants and evaluation of their potential impacts on marine organisms.

Copyright © American Institute of Chemical Engineers. All rights reserved.

387194 Detailed Thermo-Hydraulic Modelling of Crude Oil Heat Exchangers: Abnormal Fouling Behaviour Due to Inorganic Deposits

Abstract:

Crude oil fouling in heat exchangers in oil refineries leads to energy losses, operating difficulties and economic penalties. Heat exchangers at the hot end of preheat trains, where thermal fouling of organic matter is the main fouling mechanism, are traditionally considered to be more important from the point of view of energy efficiency and recovery. As a result, studies generally focus on the thermal impact of deposition of organic matter. Inorganic matter, however, is often found in analysis of deposits, although typically in heat exchangers located upstream of the desalter (cold end). Such inorganics have an important effect on the heat transfer properties of the deposit layer due to their higher conductivity with respect to that of organic deposits.

Here, a case study is presented based on recently published plant data from the Esfahan Refinery (Iran)[1]. According to the reference paper, severe fouling in the key heat exchanger, i.e. the most affected by fouling, led to inadmissible pressure drops and to the decision of shutdown, cleaning and overhaul. The analysis of the deposit inside the tubes revealed a composition with $80_{wt}\%$ inorganic matter (including SiO_2, $CaSO_4$, and iron oxides). This deposit composition is certainly unexpected, since the heat exchanger is located downstream of the desalter. This is a real example of hydraulic performance, rather than thermal, being the limiting factor for a cleaning and mitigation decision, and of abnormal fouling behaviour due to unexpected deposition of inorganics. The overhaul work, planned before the shutdown (and therefore before knowing the analysis results), was a thermal based approach aimed at reducing the high temperature of the heating fluid by introducing a new heat exchanger. Fouling mitigation was expected but, after restart, the fouling behaviour did not improve, and the pressure drop raised quickly reaching values after a year of operation of similar magnitude to those before the overhaul. Posterior analysis led to the conclusion of this high content of inorganic salts being due to desalter inefficiency.

A detailed dynamic, distributed model for heat exchangers undergoing fouling[2], previously shown to capture the complex, interacting thermo-hydraulic, fouling and ageing phenomena for thermal organic fouling, is here adapted to account for the effect of presence of inorganics. This is, to the author's knowledge, the first time this effect is included in the modelling of crude oil fouling. Pressure drop and temperature measurements, before and after a network overhaul and cleaning, and information on deposit composition are available for the tube-side of this exchanger. The detailed thermo-hydraulic model of heat exchangers, together with the available data and reasonable assumptions, is applied to simulate the unit most adversely affected by fouling. A novel modification to previous parameter estimation methodologies is introduced to fit key fouling parameters based on pressure drop measurements, instead of temperature. Information on the composition of the deposits from experimental analysis is used as an indicator of the percentage of inorganic and organic material. Then, information about the conductivity of the inorganic portion is extracted by considering possible values, corresponding to different inorganic species, and evaluating the temperature predictions against measured temperature data.

The results presented demonstrate the need of considering the enhanced conductivity of the fouling deposits due to inorganic material in order to capture main trends in reduction of thermal and hydraulic performance. They show a good match between experimental and calculated data both before and after the overhaul, despite the considerable number of assumptions, and seem to indicate that the same fouling behaviour was still present after the retrofit work. This indicates that the primary cause of fouling was not solved by the overhaul work and reinforces the hypothesis of a malfunctioning of the desalter as a possible cause.

The results also demonstrate the importance of considering both temperature and pressure drop measurements in the interpretation of plant data and fitting of model parameters, and that consideration of thermal effects in isolation would not permit to extract sufficient information about the nature of the deposit.

1. Mozdianfard MR, Behranvand E. A field study of fouling in CDU preheaters at Esfahan refinery. *Appl Therm Eng*. 2013;50(1):908–917.

2. Coletti F, Macchietto S. A Dynamic, Distributed Model of Shell-and-Tube Heat Exchangers Undergoing Crude Oil Fouling. *Ind Eng Chem Res*. 2011;50(8):4515–4533.

Acknowledgments

This research was performed under the UNIHEAT project. The authors wish to acknowledge the Skolkovo Foundation and BP for financial support, and Hexcell Ltd. for providing the modelling framework used in this work.

Copyright © American Institute of Chemical Engineers. All rights reserved.

Session: 07002 Multi-Disciplinary Engineering Analysis in Oil, Gas and Refining

AN INTEGRATED SOFTWARE FRAMEWORK FOR THE SIMULATION, MONITORING, DESIGN AND RETROFIT OF HEAT EXCHANGER NETWORKS SUBJECT TO FOULING

Francesco Coletti[1*], Sandro Macchietto[1,2]

[1] Hexxcell Ltd., Imperial College Incubator, Bessemer Building Level 2, Imperial College London, London SW7 2AZ, UK

[2] Department of Chemical Engineering, Imperial College London, South Kensington Campus, London SW7 2AZ, UK

ABSTRACT

Fouling in industrial heat exchangers is a major problem for the process industry worldwide that leads to increased operating costs, maintenance issues and environmental impact of operations. Existing software tools that assist engineers in the analysis, design, monitoring and prevention of heat exchangers and heat exchanger networks in the presence of fouling only focus on one or a subset of these aspects. They typically utilize simplifying assumptions that limit predictive capabilities within one utilization, and distinct assumptions for different applications (e.g. R&D, design, operations). As a result, work is duplicated in different engineering departments, there is no industry standard, models are not fully trusted and performance problems often arise across departments and functions. To address these problems, this paper presents a comprehensive framework that integrates the solution of monitoring, design and retrofit analysis of heat exchangers and their networks in a unified platform. At the core of the system, an advanced model allows the accurate calculations of time-varying fouling rates as a function of local conditions in the heat exchanger and fluid properties, accounting for the complex interactions between thermal and hydraulic aspects and equipment geometry.

Implemented in a user-friendly flowsheeting environment with state-of-the-art numerical solution methods and support of engineering workflows, this novel framework provides a consistent, robust and flexible system for the solution of engineering heat transfer fouling problems at all stages, from R&D to operations support.

A key benefit is the ability to consider and analyse, using consistent models and assumptions, the trade-offs between design and operational aspects. Some examples from industrial applications are presented. A second benefit is the ability of the simulation framework to work in conjunction with a state-of-the-art experimental facility for the actual measurement of crude oil fouling in conditions close to those in the refinery, utilising advanced features such as parameter estimation from dynamic data. Again, some examples are presented.

An Insight to the Fischer Tropsch Reaction

Authors: Suraj Gyawali[1], Purnima Kharidehal[1], and Daniela S. Mainardi[2]

[1]Institute for Micro-manufacturing, Louisiana Tech University
[2]Institute for Micro-manufacturing, Chemical Engineering, Louisiana Tech University

FT reaction is mainly studied and utilized to convert coal and biomass into liquid fuel. Fischer Tropsch Synthesis (FTS) is a surface catalyzed polymerization reaction in which Carbon monoxide is first adsorbed on the surface of the catalysts (mainly transition metals), and hydrogenated to form CH_x monomers. These CH_x monomers then propagates to form long chain hydrocarbons. Catalysts play an important role in product selectivity because the long chain hydrocarbons can be clean fuel, olefins, paraffin, and/or alcohol. Controlling product selectivity and overcoming catalyst deactivation has been a challenge in order to make this technology commercially available. Therefore, an optimal design of a catalyst, which can make the FTS commercially viable, includes the study of CO adsorption and the reaction mechanism of FTS on the surface of the catalyst. Similarly, the study of C-C coupling on various catalysts during the polymerization of CHx monomers will help us to determine the formation of heavy hydrocarbon, which is essential for the formation of crude oil as the product. The study of C-C coupling over catalyst surfaces will also help us to determine the surface structure of the catalysts that is needed to forward FT reaction for industrial process. Therefore, Density Functional Theory is used to study the catalytic activity in Fischer-Tropsch reaction. Alloyed nano-cluster catalysts based on Co, Fe, Ni, and Ru is being studied using Density Functional Theory. Alloyed nano-clusters are attractive because they have different catalytic properties than either of the parent metals or the flat surfaces. Moreover, composition and structure of the catalysts will have direct effect towards the activity of the Fischer Tropsch Synthesis. Cobalt based clusters with iron as core were studied using Density Functional Theory. Cobalt-Iron (Co_6Fe_6) based bimetallic cluster has shown a good result as a Fischer Tropsch Catalysts with the activation energy as low as 0.5 eV during the rate determining step of surface enol mechanisms.

389074 CFD Modeling of Bioreactor Designs

Abstract:

In recent years, there has been a growing interest in industry in biochemical processes capable of converting gaseous raw materials such as natural gas, bio-gas, or syn-gas into useful products. One of the main challenges to success in this area is the design and operation of large scale fermentation reactors that can provide for the necessary gas to liquid mass transfer while at the same time providing for the proper environment needed for a biological reaction system. As part of an ARPA-E grant to study the characteristics of different bioreactor designs, a joint project between Celanese and Calysta has been investigating the design and operating characteristics of various bioreactor designs for the utilization of methane to produce bio-based chemicals and fuels using methanotrophs. This presentation is focused on computational fluid dynamics (CFD) simulations that have been done to aid in the understanding of the benefits of different designs. The presentation will describe the modeling approach and share results from the work done to date in this effort. The objective of the overall project is to provide guidance for improved commercial bioreactor designs to further the state of the industry. More importantly, the bioreactor technologies developed in this project will position methane as a new biological feedstock for the domestic, cost-competitive production of biofuels and biochemicals.

Copyright © American Institute of Chemical Engineers. All rights reserved.

Feasibility analysis of methanol fuel processors with flexible waste heat recovery schemes

Veni Indah Christiana[a], Po-Chih Kuo[a], Wei Wu[a]*, Ya-Yan Zhou[b] and Yi-Shun Chiu[b]

[a]Department of Chemical Engineering, National Cheng Kung University, Tainan 70101, Taiwan, R.O.C.

[b]Department of Chemical and Materials Engineering, National Yunlin University of Science and Technology, Douliou, Yunlin 64002, Taiwan, R.O.C.

Abstract

The conceptual design of stand-alone methanol fuel processors using different heat recovery schemes is presented. By process optimization and simulation approaches, a flexible waste heat recovery design by utilizing the split waste heat flow is verified to reduce the waste heat loss and achieve the desired hydrogen yield. Through the operational feasibility analysis, the conflict between the optimum operating conditions and the region of feasibility is specified due to the limited waste heat from the burner and the number of degrees of freedom. Compared to a conventional methanol fuel processor using hot utilities, we find that the flexible waste heat recovery design can raise 11% hydrogen yield and reduce 3.3% waste heat loss at least.

Keywords: Methanol reforming; Feasibility; Waste heat recovery; Process design; Optimization

392616 Safe Scale-up of Energetic Processes: Recent Case Studies

Abstract:

The safe scale-up of energetic materials requires a rigorous approach to reaction hazard characterization and criticality assessment. Several recent cases where the thermal hazard criticality assessment (based on reaction calorimetry, ARC, ARSST data and resulting heat flow, MTSR,and TD24) required a process redesign or modification to safely scale-up. Our approach to process safety and details related to redesign and safe scale-up of these examples will be discussed.

Copyright © American Institute of Chemical Engineers. All rights reserved.

388819 Technology Transfer: Success Is in the Details

Abstract:

Successfully scaling technology safely from the laboratory bench to the pilot plant and from the pilot plant to a commercial plant, either owned or contracted, requires attention to detail --- not just technical detail, but planning detail and support detail.

This presentation discusses the application of dimensional analysis and the concept of the rate limiting regime to scaling technology from the labortatory to the commercial plant. This presentation also discusses the lessons learned from thirty years of scaling technology.

Copyright © American Institute of Chemical Engineers. All rights reserved.

370715 Use and Abuse of Crystallization for Process Separations

Abstract:

INTRODUCTION:

Often chemical processes require an intermediate refining step to separate reaction products and the physical properties of the components exclude typical separation unit operations such as distillation or extraction. Every effort is made to exclude crystallization because of uncertainties in the performance and/or reliability of the methodology. This presentation will address the challenges (and solutions) of integrating crystallization techniques into processes where component yield and purity are the driving forces and crystal size and shape are generally secondary concerns.

PRELIMINARY EVALUATION:

Careful evaluation of the physical and chemical properties of the proposed feed stream from which a desired component is to be recovered is essential. Reliable solubility data for two components or a phase diagram (for more than two components) is necessary for preliminary process evaluation and is best determined experimentally. MATRIC has a variety of analytical equipment and experience in obtaining and evaluating solid/liquid phase data. For a two component system MATRIC computes a chart which combines product yield and total solids in suspension, as a function of temperature and initial feed concentration. The chart basically organizes the overall material balance for the system and serves as a basis for preliminary evaluation of product yield as a function of process variables. Development and use of this chart narrows the choices available for zones of probable operation and also provides guidance regarding necessary staging of the process. A typical recovery chart will be presented and discussed.

LABORATORY TESTING:

Bench scale testing is appropriate at this stage in the process development to evaluate the crystallization behavior of the desired product (solute). A variety of techniques is available to determine the so called meta-stable or labile portion of the phase diagram. Some systems sustain limited super saturation and others may require seeding to initiate nucleation. MATRIC has equipment and experienced personnel to investigate the need for a seeding policy. A chart describing a situation where high supercooling and careful seeding are required is included. At this stage of investigation it is advisable to conduct small scale batch crystallization tests to obtain information regarding particle shape (habit). As the preliminary testing proceeds it is advisable to document the crystal habit with photomicrographs to detect any dramatic change as the process development (and impurities) change. Preliminary information regarding particle size distribution is also desirable at this stage. Generally product purity studies are a little premature at this stage; however, product recovery via filtration techniques followed by washing will give a reasonable idea of the ease or difficulty in achieving desired purity.

PILOTING:

Piloting of the proposed process serves several purposes. If the proposed process operation is batch the techniques for crystallizing, separating and washing the product can be verified on a small scale and in similar equipment to a full scale operation. Sufficient material can be produced to conduct separation and washing studies and downstream drying investigations if the product must be dried. Solid-liquid separation schemes should be considered and can be evaluated at MATRIC facilities. At this stage of process development continuous operation may be of interest and process equipment can be designed to test this concept. Often product testing by the client for acceptability is also a goal during this phase.

If particle size distribution is specified in some way studies can be conducted in a continuous mode to develop the kinetics of the system. MATRIC has piloted several systems and conducted kinetic studies.

NOTE: MATRIC also has laboratory equipment to investigate the feasibility of using so-called "Melt Crystallization" to separate and purify components. The product from this type of separation operation is a "melt" which is ideal if the product is not required to be in a particulate form.

Authors:

Charles Moyers is a senior engineering scientist at Mid-Atlantic Technology, Research and Innovative Center (P.O. Box 8396, So. Charleston, WV 25303). He is a Fellow of AIChE, a PE (WV), and has co-taught several AIChE education courses including Industrial Crystallization, Liquid-Solid Separation, and Drying. He holds a BS from Virginia Tech, an MS from West Virginia U. and a PhD from U. of Delaware, all in chemical engineering. (E-mail: charles.moyers@matricinnovates.com)

Brooke Albin (presenter) is a chemical engineer at Mid-Atlantic Technology, Research and Innovative Center (P.O. Box 8396, So. Charleston, WV 25303). She earned a BS in biology and BA in chemistry from Alderson-Broaddus College and an MSE in chemical engineering from West Virginia University. (Email: brooke.albin@matricinnovates.com)

Copyright © American Institute of Chemical Engineers. All rights reserved.

AIChE
The Global Home of Chemical Engineers

國立中央大學
National Central University

Stabilization and Spheroidization of Ammonium Nitrate: Co-crystallization with Crown Ethers and Spherical Crystallization by Solvent Screening

Chem. Eng. J. **2013**, *225*, 809–817

Department of Chemical & Materials Engineering
National Central University, Taiwan (R.O.C.)

Reporter: Chih-Wei Chen
Advisor: Prof. Tu Lee
Date: 11/20/2014
E-mail: tulee@cc.ncu.edu.tw

Bio-Inspired Materials & High Throughput Screening Laboratory

Outline

- **Introduction**
  - Propellant
  - Ammonium Nitrate (AN)
- **Results and Discussion**
  - Solvent Screening — Form Space of Ammonium Nitrate Recrystallized with Benzo-18-crown-6 (B18C6)
  - Manufacturing — 1:1 Co-crystal of AN-B18C6
  - Smoke Test
  - Burning Characteristics — Gas Generator Propellant
  - Manufacturing — Spherical agglomerates
- **Conclusions**
- **Acknowledgement**

2

Propellant

- **Propellant** — pressurized gas escape through a nozzle and create thrust.

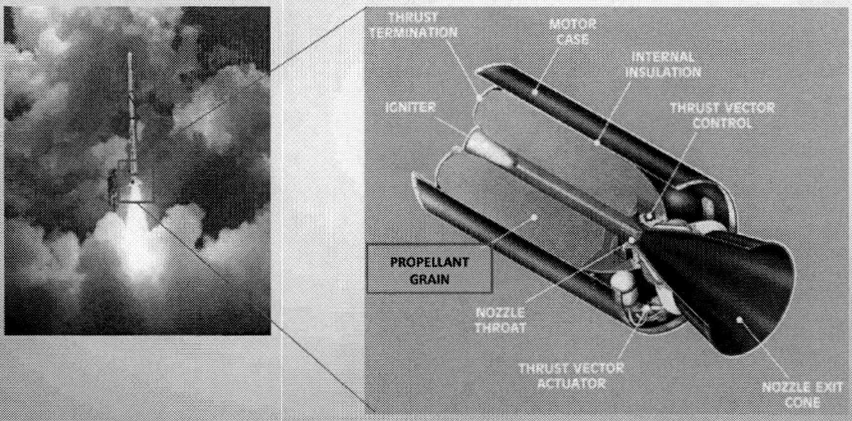

▲ Structure of rocket engine.

- **Solid-propellant gas generator** — generate electric power.

Century of Flight. http://www.century-of-flight.net

3

- **Propellant**
  - Fuel: Hydrocarbons
  - Oxidizer: Ammonium nitrate (AN)
  - Binder: Polymer
  - Catalyst: Ammonium dichromate (ADC)

High altitude — No oxygen

- **Oxygen source for burning:**

$NH_4NO_3(s) \rightarrow N_2(g) + 2H_2O(g) + \frac{1}{2}\mathbf{O_2}(g)$

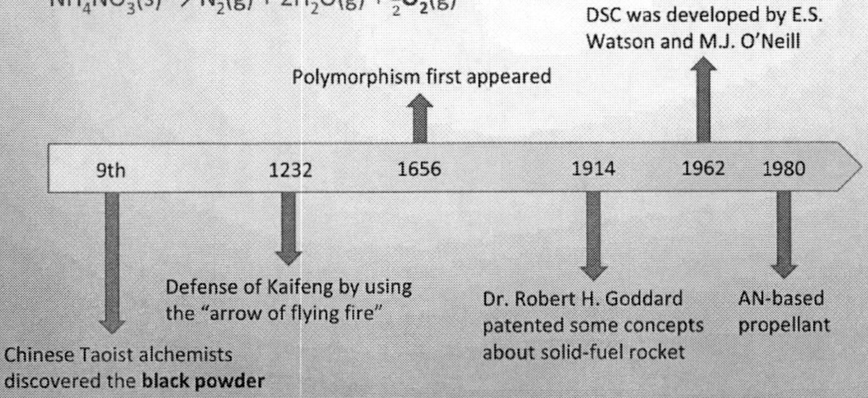

4

Ammonium nitrate

- Ammonium nitrate (NH_4NO_3):
  - ✓ **advantages**: clean burning, non-corrosive, low-hazard, low-flame temperature, inexpensive, easily available energetic material
  - X **drawbacks**: (1) polymorphic transition near ambient temperature - density(volume) change
    (2) caking and smoke generation upon burning
    (3) poor flowability for packing

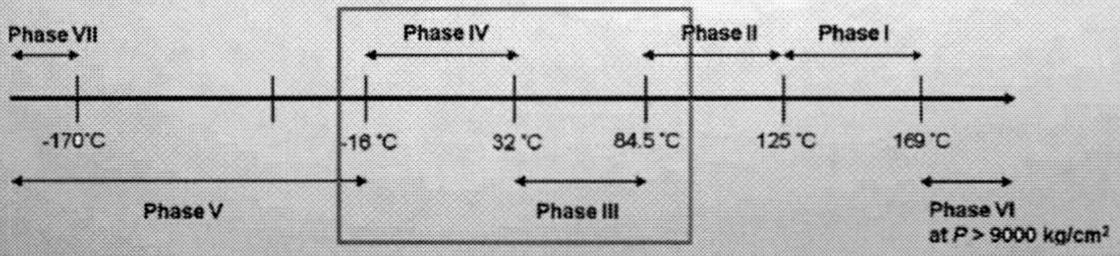

- The first two disadvantages could be minimized by forming **co-crystals.**
  The third drawback could be eliminated by forming **spherical agglomerates of co-crystals.**

Chem. Eng. J. **2013**, *225*, 809–817 5

Method

- **Co-crystallization:** chose crown ethers such as 15-crown-5 (15C5), benzo-15-crown-5 (B15C5), 18-crown-6 (18C6) and **benzo-18-crown-6 (B18C6)** as a co-former to form 1:1 co-crystals.

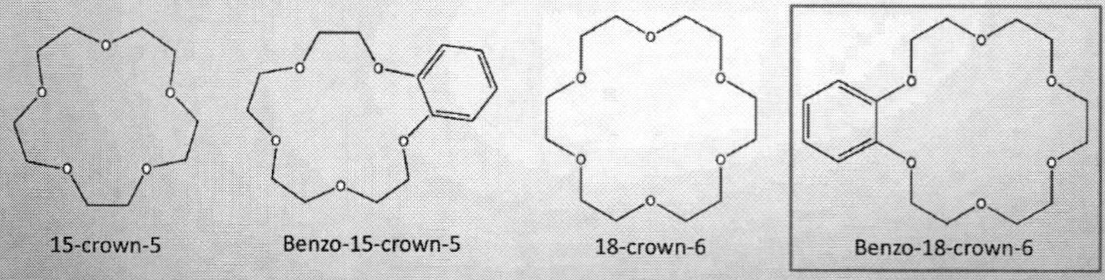

- **Spherical crystallization:** the formation of agglomerates was promoted by the addition of a small amount of the bridging liquid, which was immiscible with the antisolvent.

Chem. Eng. J. **2013**, *225*, 809–817 6

Use Test

Identification test

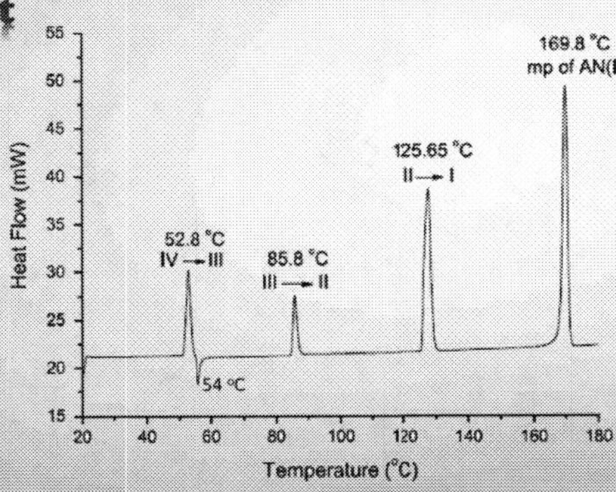

- According to differential scanning calorimetry (DSC) scan, there are **four** forms of the purchased ammonium nitrate.
- The polymorph of the purchased ammonium nitrate powders was Form IV at room temperature.

Chem. Eng. J. **2013**, *225*, 809–817 7

Solvent Screening

To construct the form space for Form IV ammonium nitrate by solvent screening

- 4 good solvents (yellow blocks): **ethanol, dimethyl sulfoxide (DMSO), methanol** and **water**.
  [Solubility > 10mg/mL]
- 18 bad solvents (red blocks) as **antisolvent** to recrystallize: *n*-heptane, xylene, *p*-xylene, **ethyl acetate**, toluene, **methyl *tert*-butyl ether (MTBE)**, benzene, methyl ethyl ketone (MEK), chloroform, tetrahydrofuran (THF), *N,N*-dimethylaniline, acetone, 1,4-dioxane, nitrobenzene, *n*-butyl alcohol, isopropyl alcohol (IPA), benzyl alcohol and acetonitrile.

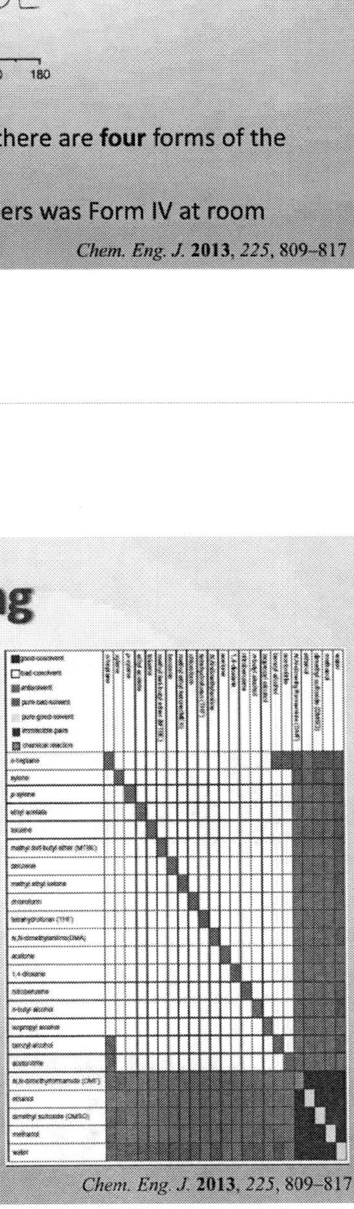

Chem. Eng. J. **2013**, *225*, 809–817 8

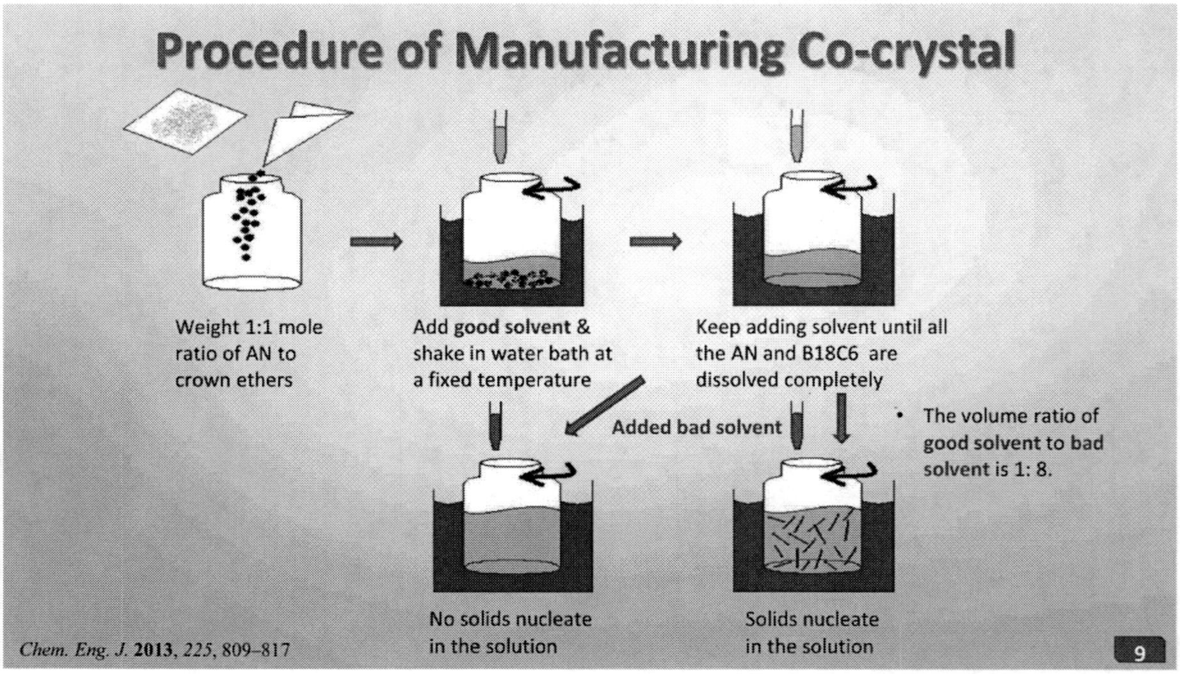
Procedure of Manufacturing Co-crystal
Weight 1:1 mole ratio of AN to crown ethers
Add good solvent & shake in water bath at a fixed temperature
Keep adding solvent until all the AN and B18C6 are dissolved completely
Added bad solvent
• The volume ratio of good solvent to bad solvent is 1: 8.
No solids nucleate in the solution
Solids nucleate in the solution
Chem. Eng. J. 2013, 225, 809–817
9

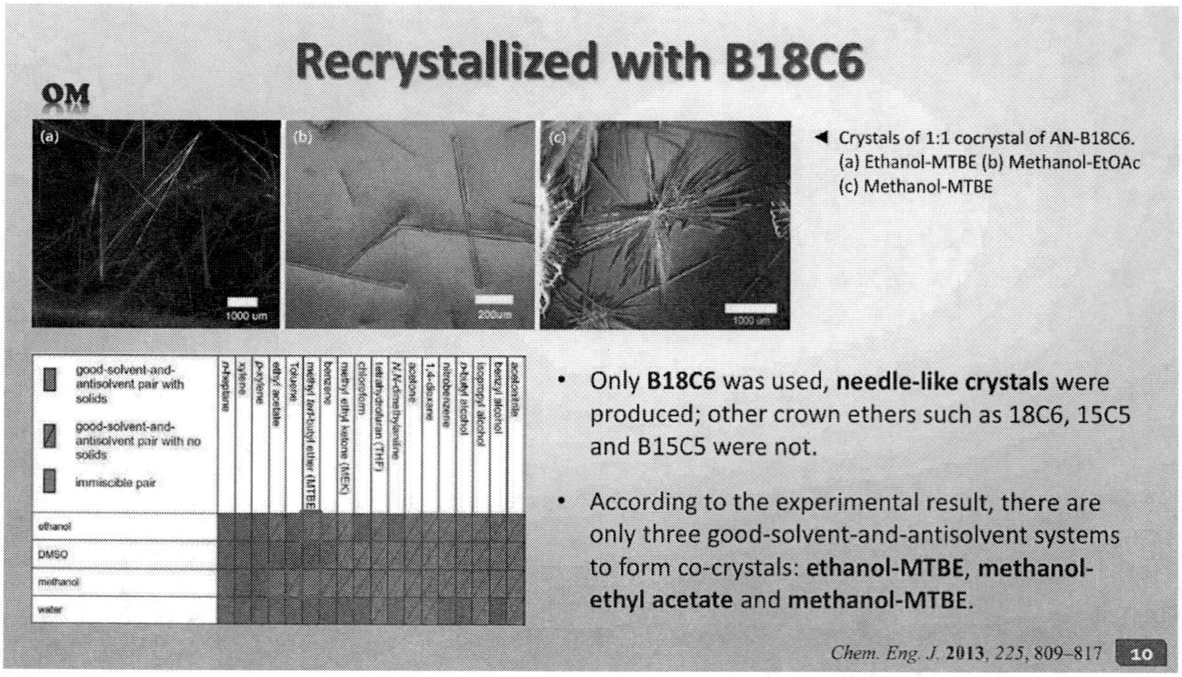
Recrystallized with B18C6
OM
(a)
(b)
(c)
1000 um
200um
1000 um
◄ Crystals of 1:1 cocrystal of AN-B18C6. (a) Ethanol-MTBE (b) Methanol-EtOAc (c) Methanol-MTBE
good-solvent-and-antisolvent pair with solids
good-solvent-and-antisolvent pair with no solids
immiscible pair
n-heptane
xylene
p-xylene
ethyl acetate
Toluene
methyl tert-butyl ether (MTBE)
benzene
methyl ethyl ketone (MEK)
chloroform
tetrahydrofuran (THF)
N,N-dimethylaniline
acetone
1,4-dioxane
nitrobenzene
n-butyl alcohol
isopropyl alcohol
benzyl alcohol
acetonitrile
ethanol
DMSO
methanol
water
• Only B18C6 was used, needle-like crystals were produced; other crown ethers such as 18C6, 15C5 and B15C5 were not.
• According to the experimental result, there are only three good-solvent-and-antisolvent systems to form co-crystals: ethanol-MTBE, methanol-ethyl acetate and methanol-MTBE.
Chem. Eng. J. 2013, 225, 809–817
10

Characterization of Co-crystal

DSC **PXRD**

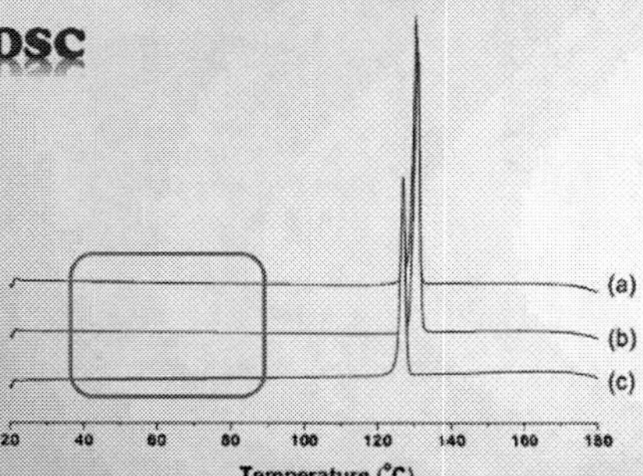

▲ Crystals of 1:1 cocrystal of AN-B18C6. (a) Methanol-MTBE (b) Methanol-EtOAc (c) Ethanol-MTBE

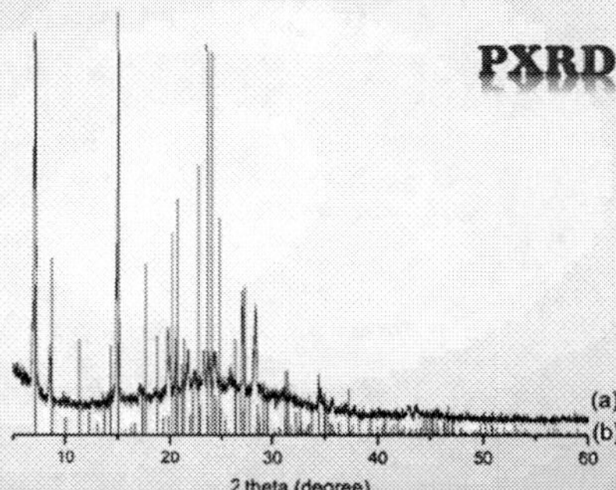

▲ (a) crystals grown from the ethanol-MTBE system (b) the calculated PXRD pattern based on SXD of crystals grown from the ethanol-MTBE system

- The DSC scans, endotherm near **125-129°C**, it illustrates that these crystals are the **same** kind; no polymorphic transition occurred between **32 to 85.5°C** .
- In addition, the good matching between the XRD diffraction diagrams of crystals validated the antisolvent precipitation method and indicated the purity of all crystals.

Chem. Eng. J. **2013**, *225*, 809–817 11

Characterization of Co-crystal

TGA

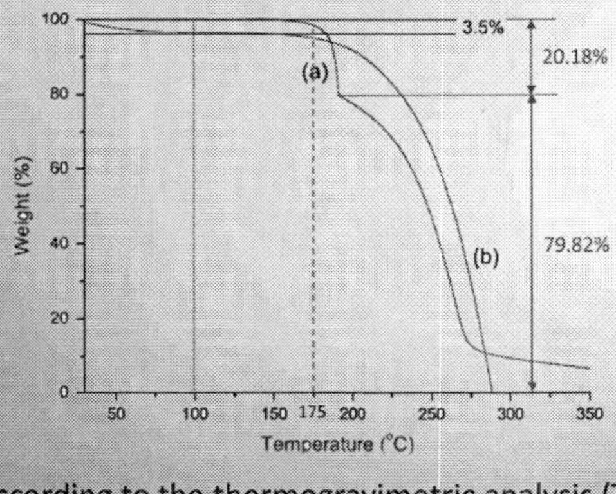

(a) crystals harvested from ethanol-MTBE system:

✓ the first weight loss at about 175°C
→ melting points = 125-129°C

✓ stoichiometric ratio:

$$\frac{\text{the first weight loss (AN)}}{\text{the second weight loss (B18C6)}} = \frac{20.18\%/80.04g \cdot mol^{-1}}{79.82\%/312.34g \cdot mol^{-1}} \approx 1$$

(b) purchased Form IV ammonium nitrate

- According to the thermogravimetric analysis (TGA) scan, there was **no significant weight loss** of 1:1 co-crystals of AN-B18C6 until 175°C, on the contrary, the purchased Form IV ammonium nitrate with **3.5 wt% weight loss from 30 to 100°C** due to dehydration. Therefore, hygroscopicity of ammonium nitrate was also minimized by co-crystallization.

Chem. Eng. J. **2013**, *225*, 809–817 12

Smoke Test

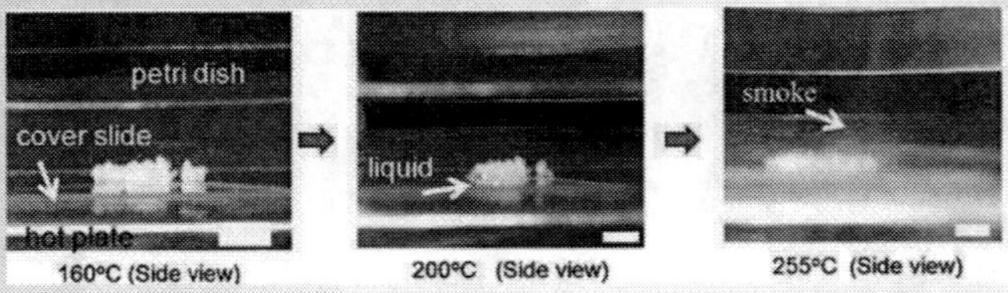

The purchased **Form IV** ammonium nitrate powders *were transformed to* **Form I** ammonium nitrate powders when heated on the hot plate in air.

- The gaseous ammonia reached cooling air and condensed to form ammonium nitrate again, which would lead to smoky appearance. $NH_4NO_3 \leftrightarrow NH_3 + HNO_3$

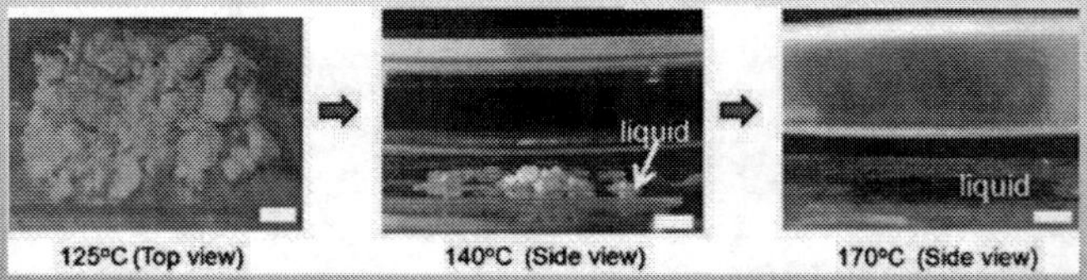

- The 1:1 co-crystals of AN-B18C6 was bonded with crown ether through **hydrogen-bonding** and was not allowed to evaporate into the gas phase to form smoke.

Chem. Eng. J. **2013**, *225*, 809–817 13

Burning Characteristics

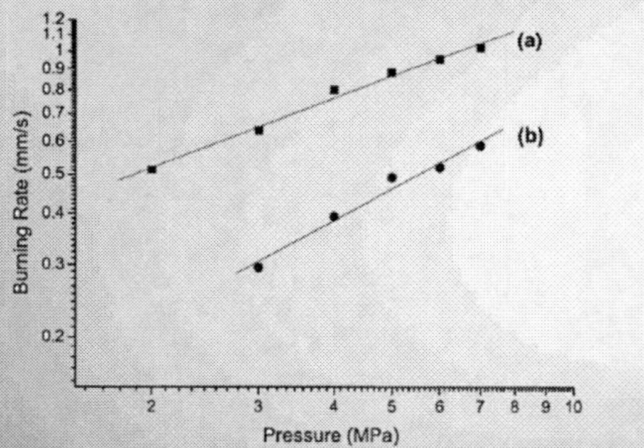

(a) AN-based propellant contained 20% **ammonium dichromate** by weight

(b) AN-based propellant contained 20% **1:1 co-crystal of AN–B18C6**

Saint-Robert's law (Veille's law)

- $r = a \cdot P^n$ ($P < 2000$ psia)
  - a is a constant
  - n is pressure exponent

a) $r = 0.35 \cdot P^{0.62}$

b) $r = 0.13 \cdot P^{0.8}$

➢ **Lower burning rate**

{ Lower flame temperature
Slower burning

- The use of lightweight hardware which would then withstand lower temperatures for a longer period.
- The utilization of an electrical generator run by hot gases given off from a slower burning rate which would in turn produce electric power for a longer time.

Chem. Eng. J. **2013**, *225*, 809–817 14

Spherical agglomerates

- **Spherical crystallization**: the formation of agglomerates was promoted by the addition of a small amount of the bridging liquid, which was **immiscible** with the **antisolvent**.
- **Water** was environmentally benign and immiscible with both ethyl acetate and MTBE, water was selected as the bridging liquid.

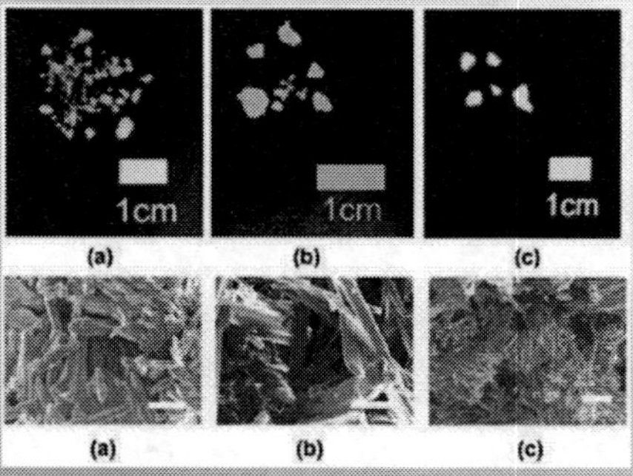

(a) 2.5 mL ethanol + 12.5 mL MTBE + 0.26 mL water
(b) 0.64 mL methanol + 11.6 mL ethyl acetate + 0.14 mL water
(c) 0.64 methanol + 11.6 mL MTBE + 0.08 mL water

	Methanol-MTBE-water	Methanol-ethyl acetate-water	Ethanol-MTBE-water
Yield (%)	87.3	24.8	52.3
Length mean diameter (cm)	0.297	0.293	0.146
Population density (1/cm^3)	20.6	10.9	1.85
Sphericity	0.86	0.82	0.71
Percentage friability index (%)	100	100	100
Angle of repose (deg.)	25.2	41.6	24.3

Methanol–MTBE–water system was the best!

- ✓ high yield
- ✓ high population density
- ✓ high sphericity
- ✓ high friability (i.e., greater strength)
- ✓ low angle of repose (i.e., good bulk flowability)

Chem. Eng. J. **2013**, *225*, 809–817 15

Conclusions

- This research had been successfully demonstrated that **co-crystallization** with crown ether could be utilized to minimize the drawbacks of ammonium nitrate:
  - ➢ polymorphic transition near room temperature
  - ➢ caking and smoke generation upon burning caused by the hygroscopic effect of ammonium nitrate
- The flowability of 1:1 co-crystals of AN-B18C6 could be further enhanced by **spherical crystallization**.

Chem. Eng. J. **2013**, *225*, 809–817 16

Acknowledgement

- My advisor – Prof. Tu Lee
- My lab partners
- My family
- National Science Council (Ministry of Science and Technology)
  - NSC 99-2119-M-008-016

17

Thanks for your attention !

18

Time To Maximum Rate Under Adiabatic Conditions Within 24hr ($TMR_{ad\ 24}$)

Christopher Morrison
Susan Shilcrat

Disclaimer:
This information is intended for educational purposes only by GlaxoSmithKline (hereinafter "GSK"). By offering this informational resource, GSK makes no representations about the suitability of this information for any purpose. The use of such information is entirely voluntary and reliance on it should only be undertaken after an independent review of it by qualified experts. In no event shall GSK be liable for any special, direct, indirect, consequential, exemplary, and/or any other damages whatsoever resulting from the use of any of this information.

Overview

- What is TMR_{ad} 24
- The Push for Change
  - 50K and 100K offset
  - Literature Review
- Calculation of TMR_{ad} 24
  - One DSC
  - One DSC + Cp
  - Three DSC + Cp
- When to Apply Each Technique

Introduction – What is TMRad 24?

- $TMR_{ad\ 24}$
  - Defined as the **T**ime to **M**aximum **R**ate given **ad**iabatic conditions such that there is **24** hours of time in order to correct the processing issue(s) prior to the reaction/decomposition reaction achieving its maximum rate of reaction
  - Based on kinetics of the reaction/decomposition reaction

Why Change? $TMR_{ad\ 24}$ vs. 100K Rule

- Even the 100K rule is unable to give an appropriate level of safety for some chemicals/mixtures when the onset temperature of decomposition is above 180°C.
- Neither the 50K nor the 100K rule take into account the kinetics of a decomposition reaction.
- $TMR_{ad\ 24}$ takes into consideration the kinetics of decomposition and in doing so may avoid the need for some ARCs for compounds with onset temperatures of decomposition below 180°C.
- Figure 11.8 from Francis Stoessel's "Thermal Safety of Chemical Processes" 2008.

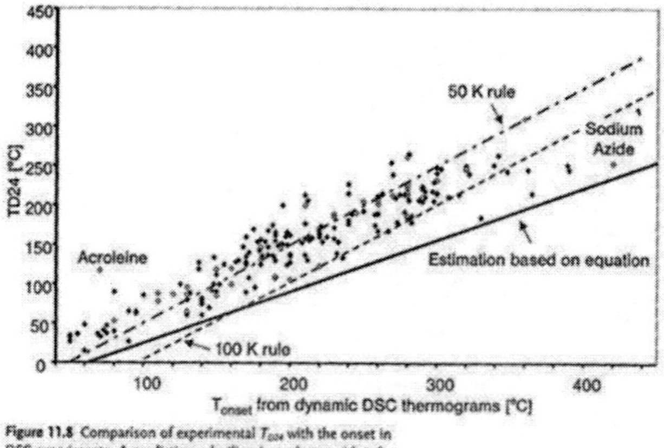

Figure 11.8 Comparison of experimental T_{D24} with the onset in DSC experiments. A prediction rule (lines) can be considered safe if all experimental values are located above the line representing the rule. White dots are pure compounds, black dots are reaction mixtures and distillation residues.

Literature Review of Energetic Materials

Source	Compound	Tonset °C	100K Rule	1 DSC TMRad 24	Tonset ARC	ARC 30K Rule	Lit. Ea (kJ/mol)	Lit. TMRad 24, °C
Bretherick's	4-Nitrobenzaldehyde	220	120	108	226	196	202	171
Bretherick's	2-Nitroaniline	280	180	150			171	223
Bretherick's	3-Nitroaniline	280	180	150			149	213
Bretherick's	4-Nitroaniline	280	180	150			160	217
Bretherick's	2-Nitrobenzoic acid	230	130	115			156	189
Bretherick's	2-Nitrophenol	260	160	136			135	182
Bretherick's	3-Nitrophenol						148	213
Bretherick's	4-Nitrophenol	240	140	122			117	185
Bretherick's	Phenyl azide	130	30	45			119	87
Bretherick's	3-Nitrobenzaldehyde	180	80	80	166	136	168	147
Bretherick's	2-Nitrobenzaldehyde	220	120	108	176	146	116	123
Bretherick's	3-Nitrobenzoic acid	320	220	178			151	245
Bretherick's	4-Nitrobenzoic acid	320	220	178			131	247
Bretherick's	2,3-Dichloroaniline						119	212
Bretherick's	2,4-Dichloroaniline						129	211
Bretherick's	2,5-Dichloroaniline						126	216
Bretherick's	2,6-Dichloroaniline						227	217
Bretherick's	3,4-Dichloroaniline						146	187
Bretherick's	1,2-Diphenylhydrazine	135	35	49			139	131
Bretherick's	4-Methoxy-1,3-phenylenediamine	260	160	136			114	207
Bretherick's	4-chloro-1,2-benzenediamine	160	60	66			140	100
Bretherick's	4-Chloroaniline	210	110	101			124	200
Bretherick's	4-Chloro-2-aminophenol						140	110
Bretherick's	Nitrobenzene	360	260	206			216	280
Bretherick's	2-Nitrotoluene						162	232
Bretherick's	4-Nitrotoluene	300	200	164			162	232

Calculating $TMR_{ad\ 24}$ Made Simple

- Step 1: Run a DSC on the sample at a rate of 2 K/min
- Step 2: Draw a baseline on the DSC data. (Accurate if done in excel since the graph can be greatly magnified.)
- Step 3: Locate the temperature when the heat produced by the sample is 10 W/kg (0.01mW/mg) above baseline this is the T_{onset}.
- Step 4: Calculate
  - Note T_{onset} is in °C

$$TMR_{ad}\ 24 = 0.71\ T_{onset\left(q(T)=10\frac{W}{kg}\right)} - 46$$

- Step 5: Understanding the Assumptions:
  - Ea= 50 kJ/mol , low activation energy
  - Zero order kinetics
  - Cp= 1.0 kJ/kg/K

Pros and Cons to Simple (One DSC) $TMR_{ad\ 24}$

- Pros:
  - Fast and conservative
  - Requires only one DSC run
  - Only requires the temperature when the self heating rate, q(T), exceeds the equipment's detection limit of ~10 W/kg above baseline
  - Reduces the need to run an ARC at lower temperatures
  - Increases safety of processes with high onset temperatures (10 W/kg detection limit)

- Cons:
  - Does not take into account actual kinetics of the decomposition; assumes a **very conservative** low activation energy, zeroth-order kinetics, and a low specific heat

Calculating $TMR_{ad\ 24}$ One DSC

- Step 1: Run a DSC on the sample at a rate of 2 K/min
- Step 2: Draw a baseline on the DSC data. (Accurate if done in excel since the graph can be greatly magnified.)
- Step 3: Locate the temperature when the heat produced by the sample is 10 W/kg (0.01mW/mg) above baseline this is the T_{onset}.
- Step 4: Locate the temperature, T_{ref}, when the heat released, q_{ref}, is 20 W/kg above baseline (beginning of exotherm to avoid autocatalytic effects).
- Step 5: Extrapolate heat, q_T

$$q_T = q_{ref} \times \exp^{\left(\frac{E_A}{R}\left(\frac{1}{T_{ref}} - \frac{1}{T}\right)\right)}$$

- Step 6: Calculate TMR with known specific heat, Cp

$$TMR_{ad} = \frac{C_p R T^2}{q_T E_A}$$

- Step 7: Understanding the Assumptions:
  - Ea= 50 kJ/mol, low activation energy
  - Zero order kinetics

Pros and Cons to One DSC $TMR_{ad\ 24}$

- Pros:
  - Fast and conservative
  - Requires only one DSC run
  - Reduces the need to run an ARC
  - Allows for the adjustment of the specific heat, Cp, which results in a more accurate estimation of the $TMR_{ad\ 24}$
- Cons:
  - Requires the temperature, T_{onset}, when the self heating rate, q(T), exceeds the equipment's detection limit of ~10 W/kg above baseline
  - Requires the temperature, T_{ref}, when the self heating rate, q_{ref}, is at ~20^{+} W/kg above baseline
  - Does not take into account actual kinetics of the decomposition; assumes a conservative low activation energy and zeroth-order kinetics

Calculating $TMR_{ad\ 24}$ Using Three DSC Experiments

- Step 1: Run three DSC on the sample at three different heating rates (1-8 K/min)
- Step 2: Draw a baseline on the DSC data.
- Step 3: Locate the temperature when the heat produced by the sample is 10 W/kg (0.01mW/mg) above baseline this is the T_{onset}. (Note that the DSC software is capable to do steps 2 and 3.)
- Step 4: Locate the temperature, T_{ref}, when the heat released, q_{ref}, is ~20 W/kg above baseline for a DSC run at 2-3 K/min.

Calculating $TMR_{ad\ 24}$ Using Three DSC Experiments

- Step 5: Enter heating rate and Onset Temperatures in the green boxes
  - Note: A correlation of 0.990 will result in an approximate TMR error of 8⁺ hrs
  - Note: A correlation of 0.999 will result in an approximate TMR error of 3 hrs

Heating rate (K/min)	LN Heating rate	Onset Temperature (°C)	Temperature (K)	1/T
1	0	213.05	486.2	0.002056767
3	1.098612289	228.48	501.63	0.001993501
5	1.609437912	236.89	510.04	0.001960631

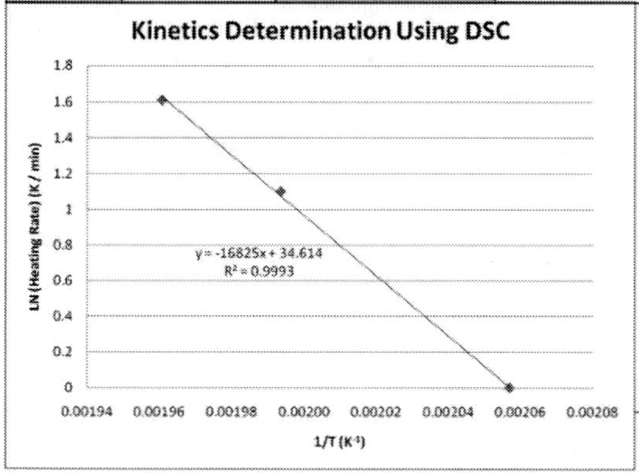

Intercept	34.61	LN(Pre-Exponent)
Pre-exponential factor	1.08E+15	A
Slope	-16824.97	-Ea/R (K)
Gas Constant	8.314	R (J/mol K)
Activation energy	139883	J/mol
Activation energy	139.88	kJ/mol

The activation energy is now known and can be applied to the decomposition reaction.

Calculating $TMR_{ad\ 24}$ Using Three DSC Experiments

- Step 6: Extrapolate the self heating rate, q_T

$$q_T = q_{ref} \times \exp^{\left(\frac{E_A}{R}\left(\frac{1}{T_{ref}} - \frac{1}{T}\right)\right)}$$

- Step 7: Calculate TMR with known specific heat, Cp, and activation energy, E_A

$$TMR_{ad} = \frac{C_p R T^2}{q_T E_A}$$

- Step 8: Understanding the Assumptions:
  - Zero order kinetics

Calculating $TMR_{ad\ 24}$ Using Three DSC Experiments

- Step 9: Do a sanity check of the data and values calculated
  - Run an isoage DSC or ARC at the temperature corresponding to a TMR_{ad} of ~8 hrs
  - Adjust TMR for Phi factor, Φ
    - Where:
    - M_c is the mass of the container
    - C_{pc} is the specific heat of the container
    - M_m is the mass of the material
    - C_{pm} is the specific heat of the material
    - TMR_{exp} is the experimental TMR
    - $TMR_{\Phi=1}$ is representative of the TMR on large scale

$$\Phi = \frac{(m_c C_{pc}) + (m_m C_{pm})}{(m_m C_{pm})}$$

$$TMR_{\Phi=1} = \frac{TMR_{\text{exp}}}{\Phi}$$

  - The TMR_{ad24} is considered conservative and safe as long as the experimental TMR_{ad8} is longer than the calculated TMR_{ad8}

Pros and Cons to Three DSC Determination of $TMR_{ad\ 24}$

- Pros:
  - Faster than an ARC and conservative
  - Reduces the need to run an ARC, can use DSC instead
  - Activation energy is known for the decomposition reaction
  - Allows for the adjustment of the specific heat, Cp, and activation energy, E_A, which results in a more accurate estimation of the $TMR_{ad\ 24}$
- Cons:
  - Requires the temperature, T_{onset}, when the self heating rate, q(T), exceeds the equipment's detection limit of ~10 W/kg above baseline
  - Requires the temperature, T_{ref}, when the self heating rate, q_{ref}, is at ~20⁺ W/kg above baseline for one of the DSC runs at 2-3 K/min
- Note: Activation energy can also be calculated from ARC exo files and the above mentioned process of using three DSCs could be used with one ARC

When to Apply Each Technique: Rules of Use

- Simplified TMR_{ad24} (One DSC)
  - Good for majority of decompositions
    - If autocatalytic may want to consider an isoage/isothermal ARC/DSC
  - If there is a melt that goes into an exotherm take T_{onset} from the beginning of the melt
- TMR_{ad24} C_p Adjusted (One DSC)
  - Allows for adjustment of Cp
  - Requires q_{ref} and T_{ref} in order to extrapolate q_T at T
  - If there is a melt that goes into an exotherm, q_{ref} and T_{ref} will not be accurate, so revert back to the simplified TMR_{ad24} calculation
- TMR_{ad24} E_A calculated and C_p Adjusted (3 DSC or 1 ARC and 1 DSC)
  - Calculation on E_A
  - Allows for adjustment of Cp
  - Requires q_{ref} and T_{ref} in order to extrapolate q_T at T
  - If there is a melt that goes into an exotherm, q_{ref} and T_{ref} will not be accurate, so revert back to the simplified TMR_{ad24} calculation
  - Run either an isothermal DSC or ARC to confirm that the TMR is conservatively accurate

Questions?

Food for Thought Regarding Heat Loss on Scale

Typical Heat Loss (Table 11.9 from Pharmaceutical Industry R&D to Manufacturing, 2011.)		
	Heat Loss (W/kg/K)	Time for 1°C Loss at 80°C
5000 L rxr	0.027	43 min
2500 L rxr	0.054	21 min
100 mL beaker	3.68	17 s
10 mL test tube	5.91	11 s
1 L Dewar	0.018	62 min

Onset Temperature from DSC (1 of 3)

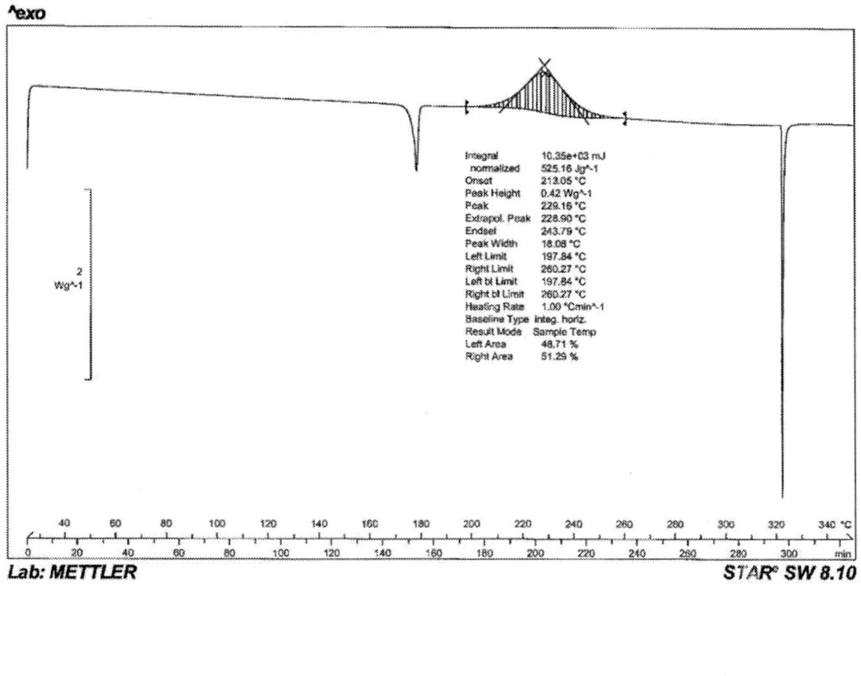

Onset Temperature from DSC (2 of 3)

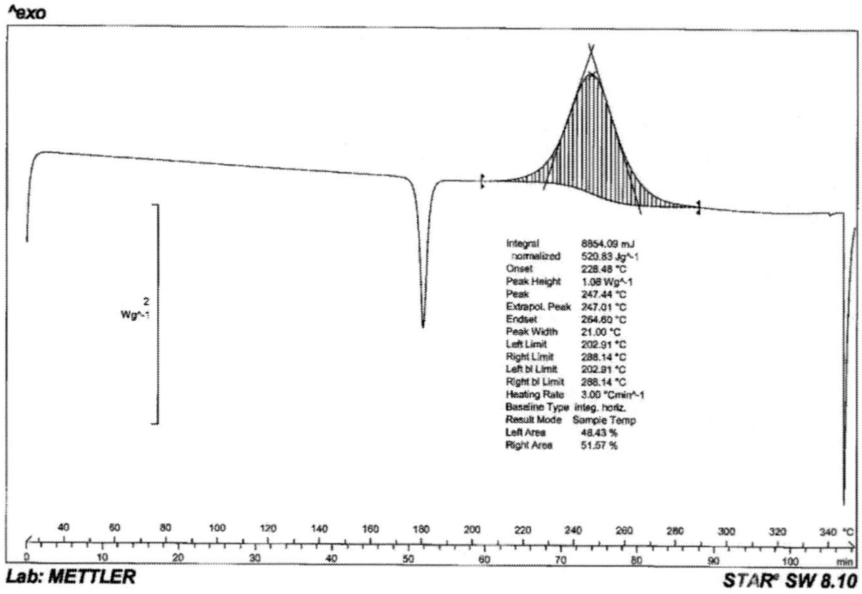

Onset Temperature from DSC (3 of 3)

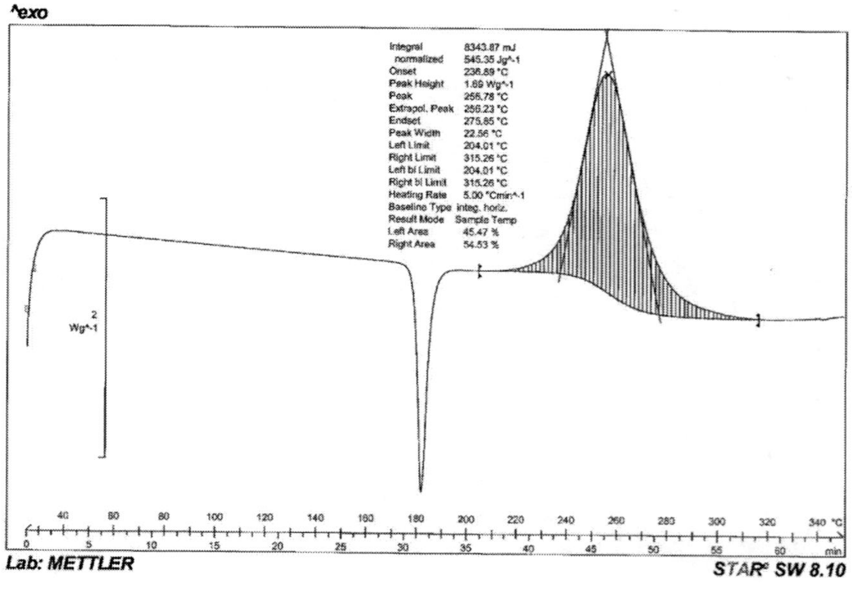

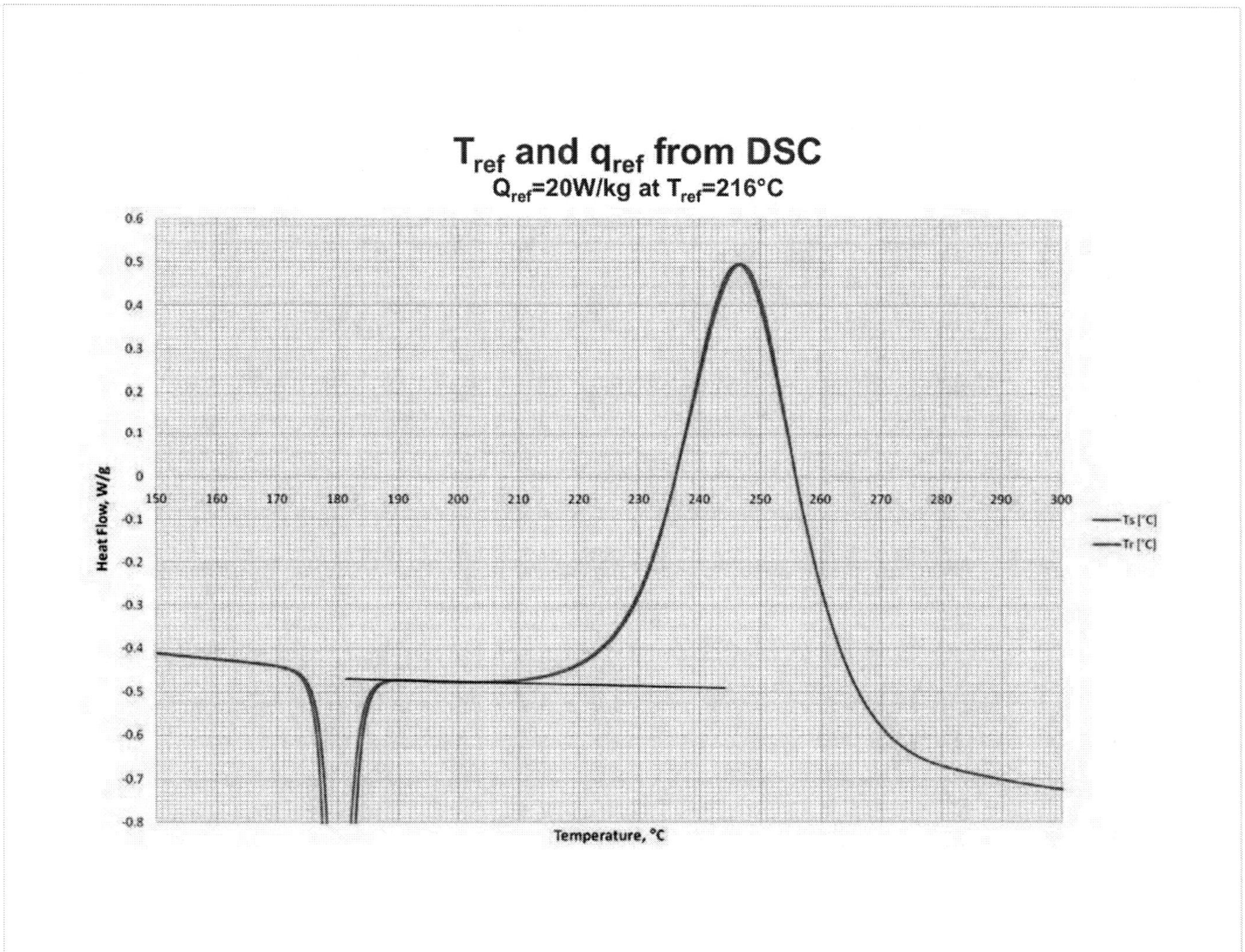
T_ref and q_ref from DSC
Q_ref=20W/kg at T_ref=216°C
Heat Flow, W/g
Temperature, °C
Ts [°C]
Tr [°C]
0.6
0.5
0.4
0.3
0.2
0.1
0
-0.1
-0.2
-0.3
-0.4
-0.5
-0.6
-0.7
-0.8
150
160
170
180
190
200
210
220
230
240
250
260
270
280
290
300

Abstract

Fouling is a complex phenomenon which commonly occurs in crude oil heat transfer equipment, reducing heat transfer, increasing pressure drops and in general resulting in reduced energy recovery efficiency and increased capital and operating costs of refineries.

Measuring with accuracy the fouling behaviour of an oil at or close to industrial conditions is not easy. A High Pressure Oil Rig (HIPOR), designed to overcome current limitations and operate in a regime close to refinery shell-and-tube exchangers, has been developed at Imperial College London under the direction of Prof. G.F. Hewitt. The apparatus includes a tubular and annulus flow sections with heat input supplied via Joule heating (Macchietto et al, 2011), and an experimental programme is under way.

The aim of this paper is two-fold: First, a predictive model of the HIPOR test section is briefly presented. The model is intended to assist in the analysis and interpretation of the experimental fouling data collected, and also in the development of the experimental programme itself. Second, some results are reported in respect of the calibration of the model to the equipment in clean (no-fouling) conditions, so as to establish a properly validated performance of the equipment (sensors and controllers, as well as thermo-hydraulic behaviour) in the absence of fouling and a reliable base-line for the assessment of differences due to fouling.

For the first objective, a dynamic, distributed thermo-hydraulic model of the tubular and annulus flow sections of HIPOR is presented. The model, an extension of Hexxcell's model for HiPOR, considers the two dimensional (axial and radial) heat transfer in different domains with variable thermal conductivity (heat generation), boundary conditions and other aspects (for example, heat losses) which are carefully determined. The calibration and validation of various model components through a variety of tests is described. The model predicts temperature profile and pressure drop of fluid in the test section as well as local conditions of the physical plant and fluid properties.

Instrumentation for temperature, flow rate and differential pressure measurement has been installed to record primary data for validation purposes. Heat input (heat flux) supplied via Joule heating is also carefully measured.

Tests were performed for varying inlet fluid temperatures and step changes in the heat input, enabling experimental results and model predictions to be compared. The simulation results (*temperature profile, wall temperature, differential pressure and temperature difference, outlet and inlet temperature*) show a consistent agreement with primary measurements from the rig at different process conditions, including transient responses. Furthermore, results also demonstrate the sensitivity of operating process parameters to differential pressure.

It is concluded that the model of the HIPOR rig at 'clean' oil conditions was successfully validated and could be a used as a reliable reference base to assist in the interpretation and validation of crude oil fouling data from HIPOR.

References

S. Macchietto, G.F. Hewitt, F. Coletti, B.D. Crittenden, D.R. Dugwell, A. Galindo, G. Jackson, R. Kandiyoti, S.G. Kazarian, P.F. Luckham, O.K. Matar, M. Millan-Agorio, E.A. Müller, W. Paterson, S.J. Pugh, S.M. Richardson, D.I. Wilson, "Fouling in crude oil preheat trains: a systematic solution to an old problem", *Heat Transfer Engineering*, **32**, 3, 197-215 (2011).

F. Coletti and S. Macchietto A dynamic, distributed model of shell-and-tube heat exchangers undergoing crude oil fouling. *Ind. Eng. Chem. Res.* **50** (8), pp 4515–4533 (2011).

AUTHOR INDEX

AUTHOR INDEX

AIChE
120 Wall Street, FL 23
New York, NY 10005-4020

ISBN 978-1-5108-1275-8

9 781510 812758

CURRAN ASSOCIATES INC.
proceedings.com